AF601496

Synthesis Lectures on Renewable Energy Technologies

The series, Synthesis Lectures on Renewable Energy Technologies publishes concise books, focused on technologies that harness energy from naturally occurring sources, such as sunlight, wind, water, geothermal heat, and biofuels from organic materials. These renewable energy technologies play a crucial role in transitioning away from fossil fuels, helping to mitigate the effects of climate change, and promoting a sustainable energy supply.

Yujie Yuan · Lai-Chang Zhang · Zhuo Feng

Natural Hydrogen

Geological Formation, Accumulation, and Resource Potential for a Low Carbon Future

Yujie Yuan
School of Engineering
Edith Cowan University
Perth, WA, Australia

Lai-Chang Zhang
School of Engineering
Edith Cowan University
Perth, WA, Australia

Zhuo Feng
Institute of Palaeontology
Yunnan University
Kunming, Yunnan, China

ISSN 2690-5000 ISSN 2690-5019 (electronic)
Synthesis Lectures on Renewable Energy Technologies
ISBN 978-3-032-14468-3 ISBN 978-3-032-14469-0 (eBook)
https://doi.org/10.1007/978-3-032-14469-0

This Springer imprint is published by the registered company Springer Nature Switzerland AG
The registered company address is: Gewerbestrasse 11, 6330 Cham, Switzerland

Foreword

The growing demand for low-carbon and sustainable energy has renewed scientific interest in geological hydrogen as a naturally occurring resource generated by deep Earth processes. Unlike conventional hydrocarbons, natural hydrogen is not confined to sedimentary basins but is closely associated with tectonic activity, fluid–rock interactions, and crust–mantle processes in deep earth. Its occurrence challenges traditional resource concepts and calls for a reassessment of subsurface energy systems from a geological perspective.

This book provides a systematic synthesis of current understanding of the generation, migration, and accumulation of natural hydrogen in diverse geological settings. Drawing on advances in structural geology, geochemistry, petrology, and hydrogeology, it establishes a coherent framework for interpreting hydrogen occurrences worldwide. By comparing hydrogen systems with conventional oil and gas systems, the volume highlights fundamental differences in driving mechanisms, transport behavior, and storage potential, thereby clarifying key scientific and exploration challenges.

In addition to summarizing established knowledge, this book identifies critical uncertainties and emerging research frontiers, including transient accumulation, rapid diffusion, and the balance between continuous generation and dissipation. It is expected to serve as a valuable reference for researchers and graduate students in Earth sciences, as well as for professionals engaged in energy resource assessment and sustainable development. By strengthening the geological basis of natural hydrogen studies, this volume aims to support future exploration efforts and contribute to the broader transition toward diversified clean energy resources.

Professor Ping'an Peng
Member of the Chinese Academy of Sciences
Guangzhou Institute of Geochemistry, Chinese Academy of Sciences

Preface

Hydrogen is the most abundant element in the universe, however only in recent years has 'natural hydrogen' (the molecular hydrogen generated and stored within the Earth's crust) emerged as a subject of renewed scientific and industrial interest. While hydrogen has long been central to discussions of the clean energy transition, most efforts have focused on manufactured hydrogen, produced through electrolysis or hydrocarbon reforming processes. In contrast, natural hydrogen represents a potentially transformative primary energy resource: it is continuously generated by geological processes, potentially renewable on human timescales, and with a remarkably low carbon footprint.

This book, *Natural Hydrogen: Geological Formation, Accumulation, and Resource Potential for a Low Carbon Future*, explores this emerging frontier at the intersection of geoscience, geochemistry, and energy resource engineering. It aims to bridge the knowledge gap between fundamental molecular-scale process and large-scale geological accumulation systems. Drawing upon recent advances in geological genesis, accumulation mechanism, and field discovery, the book examines key hydrogen generation pathways, including water–rock reactions, radiolysis, serpentinization, and mantle degassing, as well as migration dynamics, trapping mechanisms, and subsurface resource distributions.

The objectives of this book are twofold. First, it seeks to provide a coherent scientific framework for understanding how natural hydrogen forms and evolves within the deep Earth system. Second, it evaluates the potential of natural hydrogen as a viable and sustainable energy resource in the context of a global transition toward net-zero carbon future. As with any emerging discipline, substantial uncertainties remain, ranging from the kinetics and longevity of hydrogen production to the geophysical and geochemical signatures that reveal its presence in the subsurface. Nevertheless, the pace of discovery is accelerating, and natural hydrogen research is increasingly positioned to contribute meaningfully to future energy systems.

This book is intended for students, researchers, and industry professionals with interests in hydrogen geoscience and engineering, subsurface rock-fluid systems, and sustainable resource development. We hope that the perspectives and insights presented here will foster further interdisciplinary collaboration, encourage exploration and innovation, and stimulate new ways of thinking how the Earth itself may support the hydrogen economy of the future. At the same time, it invites a re-examination of sustainable energy development that aligns resource utilization with environmental, social, and economic sustainability. In this context, natural hydrogen represents not only an emerging energy opportunity, but also a framework for rethinking humanity's stewardship of the planet.

Competing Interests The authors have no competing interests to declare that are relevant to the content of this manuscript.

Perth, Australia Yujie Yuan
Perth, Australia Lai-Chang Zhang
Kunming, China Zhuo Feng

Acknowledgements

We wish to extend our sincere gratitude to many individuals and institutions whose contributions, support, and inspiration were integral to the realization of this book.

First and foremost, we acknowledge the pioneering researchers whose foundational work in hydrogen geoscience, geochemistry, structural geology, and subsurface rock-fluid systems has laid the scientific framework upon which this book is built. Their field observations, experimental studies, and theoretical insights have collectively advanced the understanding of natural hydrogen generation, migration, accumulation and resource distribution, rendering this synthesis both timely and feasible.

We are deeply grateful to colleagues and collaborators from academic institutions and industrial organizations worldwide for their valuable discussions, constructive feedback, and the vibrant exchange of ideas that have helped shape the perspectives presented herein. Many of these interactions, facilitated through collaborative projects, conferences, workshops, or informal discussions, have been instrumental in refining the concepts and interpretations developed throughout the book.

We also acknowledge the support provided by research funding agencies and institutional programs that have enabled sustained investigation into subsurface hydrogen systems and related geological processes. Their unwavering commitment to fundamental and applied geoscience research has played a critical role in fostering innovation within this emerging field.

Particular thanks are extended to the students and early-career researchers whose intellectual curiosity, critical thinking, and enthusiasm have continually challenged established paradigms and inspired new lines of inquiry. Their engagement underscores the vital importance of education and mentorship in advancing interdisciplinary research at the interface of Earth science and energy studies.

The authors are grateful to the editorial and production teams for their professionalism, patience, and guidance throughout the preparation of this manuscript. Their careful handling of the review, revision, and publication processes has greatly improved the clarity and presentation of the final work.

Contents

About the Authors

Yujie Yuan is a Lecturer and the Discipline Coordinator of the Chemical, Energy, and Resources Discipline in the School of Engineering at Edith Cowan University, Australia. She received Ph.D. degree in the Western Australian School of Mines at Curtin University, Western Australia. She has served in a range of international academic roles, including Principal Chair of international conference (e.g., Chief Committee Chair of the *Intelligent Technology for Power and Energy Conference* 2025 in Amsterdam, the Netherlands), Editorial Board Members for several peer-reviewed journals (including *Advances in Geo-Energy Research, International Journal of Coal Science & Technology, Advanced Light Materials, and Scientific Reports*), the Editor of academic books and special issues. She has also been an invited keynote speaker at international conferences and has served as an honorary consultant and guest lecturer in both academia and industry. Her research focuses on low-carbon energy systems and energy storage, including carbon storage, underground hydrogen storage, natural hydrogen, shale gas/oil). Over the past seven years, she has published more than 60 peer-reviewed journal articles and book chapters. Several of her publications have been recognized as Highly Cited Papers by Web of

Science, reflecting their significant impact on the field.

Lai-Chang Zhang is a Professor and the Head of the Centre for Advanced Materials and Manufacturing at Edith Cowan University. His research interests include metal additive manufacturing, metallic biomaterials, light alloys & structures, and high-strength alloys and composites. He is a Highly Cited Researcher (Clarivate) and an Advanced Materials Laureate (IAAM, Sweden), with approximately 410 journal articles, an H-index of 95 and more than 33000 citations. He serves as Editor-in-Chief of Advanced Light Materials, Editor or Editorial Board Members for 10+ journals, e.g., International Journal of Extreme Manufacturing, Advanced Powder Materials, Advanced Engineering Materials, Scientific Reports, Acta Metallurgica Sinica (English Letters), etc.

Zhuo Feng holds the title of Donglu Distinguished Professor at Yunnan University, China, where he serves as Director of both the Institute of Palaeontology and the Yunnan Key Laboratory of Earth System Sciences. He earned his Ph.D. from the Nanjing Institute of Geology and Palaeontology, Chinese Academy of Sciences, and was awarded the prestigious Chinese National Science Fund for Distinguished Young Scholars. His research mainly focuses on plant evolution, plant–insect coevolution, and the development of terrestrial ecosystems. He has authored over 100 papers in high-impact interactional journals, including PNAS, Nature Communications, Science Advances, National Science Review, Current Biology, and Earth-Science Reviews.

List of Figures

List of Tables

1 Introduction

The development of natural hydrogen (also referred to as geological hydrogen or white hydrogen) has attracted growing scientific and industrial interest as a potentially substantial and sustainable energy resource due to its clean combustion profile and natural occurrence in geological systems [1]. Hydrogen is progressively recognized as a critical energy vector during the transition to sustainable energy frameworks. As an environmentally benign energy vector, H_2 possesses the capacity to substantially diminish greenhouse gas emissions when utilized as a fuel [2]. Its classification into three distinct categories, green, blue, and white, reflects diverse production methodologies, each characterized by unique processes and environmental footprints (Fig. 1.1). Green hydrogen, synthesized via water electrolysis, leverages renewable energy sources such as wind, solar, or hydropower to dissociate water into H_2 and O_2 [3]. This method facilitates carbon–neutral hydrogen generation, positioning it as a cornerstone of sustainable energy strategies [4]. Conversely, blue hydrogen represents a low-carbon derivative of natural gas, synthesized through steam methane reforming or autothermal reforming.By coupling these processes with carbon capture and storage (CCS), it is possible to effectively reduce CO2 emissions, thereby minimizing the overall environmental impact of the production cycle [5]. This hybrid approach offers a transitional solution to decarbonize conventional fossil fuel-based hydrogen synthesis. Meanwhile, white hydrogen, or geologically occurring hydrogen, is a naturally formed variant that accumulates within subsurface systems and can be extracted through direct exploration techniques, eliminating the requirement for anthropogenic synthesis [1]. Unlike H_2 produced from water electrolysis or methane reforming, natural hydrogen (white hydrogen) is found in subsurface geological formations. The advancement of natural hydrogen systems could significantly transform the

Y. Yuan et al., *Natural Hydrogen*, Synthesis Lectures on Renewable Energy Technologies, https://doi.org/10.1007/978-3-032-14469-0_1

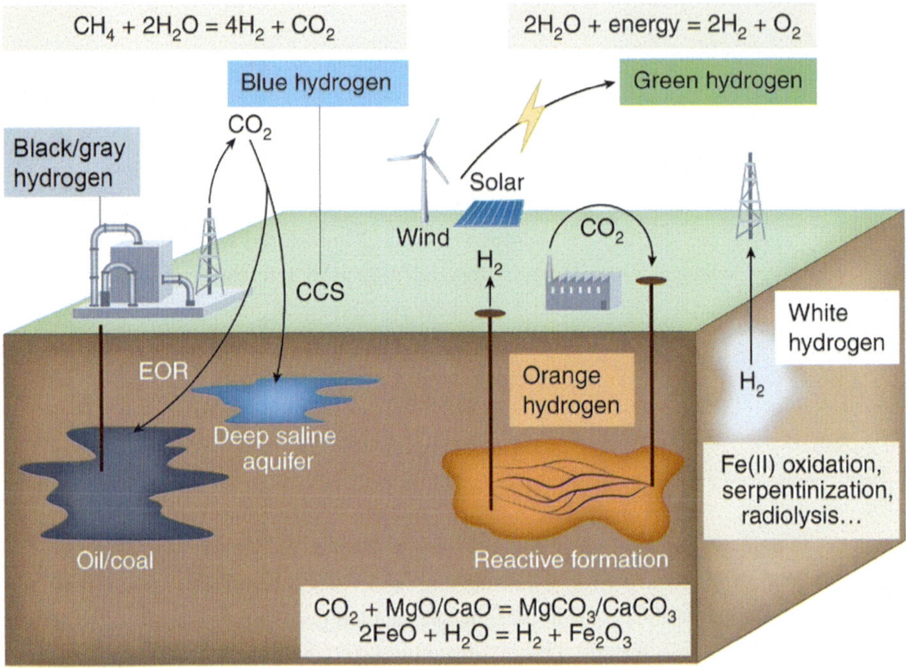

Fig. 1.1 Different methods of hydrogen production [3]

hydrogen economy by offering an environmentally sustainable and low-carbon energy alternative [6].

1.1 Natural Hydrogen System

Natural hydrogen accumulations (white hydrogen), constitute a promising yet under-exploited energy resource [1]. Unlike green or blue hydrogen, which rely on energy-demanding synthetic methods, white hydrogen is generated and stored naturally within the Earth's crust through distinct geological processes. A primary mechanism is serpentinization, a reaction in which ultramafic rocks (e.g., peridotite) undergo chemical alteration upon contact with water at elevated temperatures, releasing H_2 as a secondary product [7]. These reactions are common in geologically active settings such as mid-ocean ridges and subduction zones, which are characterized by prolonged water–rock interactions [8]. Additionally, radiolytic decomposition serves as a critical source of natural hydrogen, driven by ionizing radiation from uranium, thorium, and other radioactive elements that dissociate water molecules into H_2 and O_2 [9]. This process dominates in stable cratonic regions like Precambrian shields, where prolonged geological stability allows radioactive materials to persist and drive radiolysis over millennia [10].

Geological hydrogen systems predominantly occur in distinct geological settings, including ultramafic lithologies, crystalline basement structures, and sedimentary basins hosting organic-rich strata [3]. These formations function as persistent subsurface storage systems, where H_2 accumulation occurs within confined reservoirs that may exhibit gradual migration toward the surface or remain preserved in structural traps [1]. The utilization of these geological hydrogen systems presents a novel prospect for direct exploitation, offering enhanced energy efficiency relative to conventional synthetic production techniques.

Historically, natural hydrogen was largely overlooked due to the dominance of fossil fuel exploration and a lack of awareness regarding its potential [11–1314]. However, in the past decade, a growing number of studies and pilot projects have explored its feasibility as a clean energy source. Several key milestones highlight the current state of natural hydrogen development: (i) Hydrogen Seeps Discovery: Natural hydrogen seeps have been documented in numerous geological settings worldwide, especially within geodynamically active regions, including mid-ocean ridges, continental rift systems, and stable cratonic areas. Notable discoveries include hydrogen-rich gas seeps in Russia [15], Mali of Africa [16], the United States of America [15], and Australia [17]. (ii) Pilot Projects: In countries like Mali and Australia, pilot drilling projects have begun to evaluate the commercial viability of natural hydrogen extraction. One example is the Bourakebougou field in Mali, where hydrogen concentrations as high as 98% have been found [18]. These early initiatives are laying the groundwork for larger-scale projects. (iii) Increased Investment and Interest: With the global push for decarbonization, natural hydrogen is attracting interest from both governments and the private sector. Companies involved in oil and gas exploration are beginning to pivot toward hydrogen exploration, while governments are funding research to assess natural hydrogen's potential contribution to energy transitions [19].

1.2 Significance and Objectives

The study of natural hydrogen system is becoming a pivotal component in facilitating the shift toward sustainable energy systems. These subsurface accumulations present a distinct advantage by supplying H_2 in its native form, thereby eliminating the need for energy-intensive production processes required by conventional methods [20]. For example, (i) Clean Energy Source: H_2 combustion produces only water, making it an ideal fuel for reducing carbon emissions and meeting global decarbonization goals [21]. (ii) Abundance: While current natural hydrogen system is underexplored compared to oil and gas system, there is significant potential for large-scale deposits, especially in regions with high tectonic or volcanic activity [1, 22]. (iii) Renewability: Unlike fossil fuels,

H_2 production through geological processes could be semi-renewable, with continuous production occurring in active geological formations [23].

Unlike green hydrogen, which demands significant electrical energy for electrolysis-based production, or blue hydrogen, which relies on sophisticated carbon capture systems, natural hydrogen requires considerably less energy for extraction [24], thereby lowering both production costs and associated energy expenditures. This positions natural hydrogen as an especially promising alternative in regions with constrained access to renewable energy infrastructure or areas that possess suitable geological formations for hydrogen accumulation.

A key advantage of natural hydrogen lies in its significantly reduced environmental footprint during extraction processes. Unlike traditional fossil fuel extraction, accessing natural hydrogen system generally requires reduced surface infrastructure and employs minimally invasive drilling methods [1, 25]. This characteristic enhances its feasibility for deployment in ecologically fragile zones or locations prioritizing limited landscape disruption [26]. Furthermore, since natural hydrogen utilization does not rely on hydrocarbon combustion, its application as a fuel generates zero CO_2 emissions, establishing it as a carbon–neutral energy resource following capture and implementation [27].

Economically, natural hydrogen serves as a critical asset in stabilizing global hydrogen markets by providing a more reliable and cost-competitive supply alternative. With rising global demand for hydrogen, the utilization of natural hydrogen system could help mitigate supply–demand imbalances and price fluctuations [1]. Geologically advantaged regions, such as areas within Africa, Australia, and North America, demonstrate particular promise [28], where continuous H_2 generation through natural processes like serpentinization could establish a sustainable supply chain to complement conventional production techniques.

Beyond their economic and environmental advantages, natural hydrogen system holds strategic value through its role in enhancing energy security. Exploitation of domestic deposits enables nations to diminish dependence on imported hydrogen and carbon-based energy sources, strengthening national energy self-sufficiency [29]. The inherent geological stability demonstrated by reservoir formations, including those within ancient crystalline shields and sedimentary basins, assures their viability as durable energy storage systems that can mitigate imbalances between energy production and consumption [30].

The sustainable exploitation of natural hydrogen reserves necessitates an integrated strategy that merges geological investigations, geochemical assessments, and state-of-the-art drilling technologies [1]. Indispensable geophysical methodologies, including seismic and gravity surveys, play a pivotal role in delineating subsurface structures to detect hydrogen-rich zones [31]. Seismic imaging delivers vital insights into the depth and structural characteristics of prospective reservoirs, whereas gravity anomaly mapping identifies regions exhibiting anomalous rock density signatures that may signify H_2 occurrence [32].

These integrated approaches direct drilling operations, facilitating the strategic selection of extraction targets to reduce both financial expenditures and ecological footprint.

To achieve the seamless incorporation of natural hydrogen system into the global energy framework, sustained investment in research, exploratory initiatives, and infrastructure modernization remains imperative. By expanding upon the current scientific understanding and harnessing cutting-edge innovations, the full potential of natural hydrogen can be realized, delivering a clean, renewable energy resource that supports global decarbonization efforts [27].

This book aims to: (i) investigate the geological mechanisms governing the generation, migration, and entrapment of natural hydrogen across diverse subsurface geological settings. (ii) assess the feasibility of ultramafic rock formations acting as primary systems for H_2 generation via serpentinization reactions. (iii) examine the viability of natural hydrogen deposits as a cost-competitive and environmentally sustainable substitute for conventional hydrogen synthesis techniques. (iv) evaluate critical technical barriers in drilling, extraction, storage, and transportation systems required to harness H_2 from subsurface systems (v) establish proactive risk assessment and preventive strategies addressing H_2 leakage, operational hazards, and ecological impacts to ensure safe and sustainable extraction practices. (vi) advance scientific knowledge on natural hydrogen systems and their strategic implications in accelerating global adoption of decarbonized energy systems.

References

1. Zgonnik, V. 2020. The occurrence and geoscience of natural hydrogen: A comprehensive review. *Earth-Science Reviews* 203: 103140.
2. Ball, M., and M. Wietschel. 2009. The future of hydrogen—Opportunities and challenges. *International Journal of Hydrogen Energy* 34 (2): 615–627.
3. Osselin, F., et al. 2022. Orange hydrogen is the new green. *Nature Geoscience* 15 (10): 765–769.
4. Du, Z.M., et al. 2021. A review of hydrogen purification technologies for fuel cell vehicles. Catalysts 11 (3).
5. Blok, K., et al. 1997. Hydrogen production from natural gas, sequestration of recovered CO_2 in depleted gas wells and enhanced natural gas recovery. Energy.
6. Tasleem, S., and E. H. Alsharaeh. 2025. Role of green, yellow, blue, white and gold hydrogen in fuelling the path to net zero and sustainable future—A review. *Energy Conversion and Management* 326: 119500.
7. Sleep, N. H., et al. 2004. H_2-rich fluids from serpentinization: Geochemical and biotic implications. *Proceedings of the National Academy of Sciences of the United States of America* 101 (35): 12818–12823.
8. Allen, D. E., and W. E. Seyfried. 2004. Serpentinization and heat generation: Constraints from Lost City and Rainbow hydrothermal systems. *Geochimica Et Cosmochimica Acta* 68 (6): 1347–1354.
9. Lin, L.H., et al. 2005. Radiolytic H_2 in continental crust: Nuclear power for deep subsurface microbial communities. *Geochemistry Geophysics Geosystems* 6.

10. Bourdet, J., et al. 2023. Natural hydrogen in low temperature geofluids in a Precambrian granite, South Australia. Implications for hydrogen generation and movement in the upper crust. *Chemical Geology* 638.
11. Yuan, Y., Wu, S., Al-Khdheeawi, E.A., Lin, Z., Wang, J., Wang, M., Zhou, Y., Zhang, L.-C., Feng, Z., 2026. Sweet spot development in heterogeneous shales of lower Silurian. International Journal of Coal Geology, 313, 104902
12. Yuan, Y., Zhou, Y., Al-Khdheeawi, E., Lin, Z., You, Z., Zou, J., Iglauer, S., Feng, Z., Zhang, L.-C., 2025. Deep Shale of Lower Cambrian: Implications for Prospective Shale Gas Resources, Energy & Fuels, 39, 38
13. Yuan, Y., Wu, S., Al-Khdheeawi, E.A., Tan, J., Feng, Z., Rezaee, R., You, Z., Jiang, H., Wang, J., Iglauer, S., 2024. Substantial gas enrichment in shales influenced by volcanism during the Ordovician–Silurian transition. International Journal of Coal Geology, 295, 104638
14. Hand, E. 2023. Hidden hydrogen: Does earth hold vast stores of a renewable, carbon-free? *Science* 379 (6633): 630–636.
15. Larin, N., et al. 2015. Natural molecular hydrogen seepage associated with surficial, rounded depressions on the European craton in Russia. *Natural Resources Research* 24 (3): 369–383.
16. Maiga, O., et al. 2023. Characterization of the spontaneously recharging natural hydrogen reservoirs of Bourakebougou in Mali. *Scientific reports* 13 (1): 11876–11876.
17. Boreham, C.J., et al. 2021. Hydrogen in Australian natural gas: Occurrences, sources and resources. *The Australian Energy Producers Journal* 61 (1): 163–191.
18. Su, Y., et al. 2024. Natural hydrogen exploration: A case study of hydrogen wells in the Mali gas field in Africa and global advances (in Chinese). *Oil & Gas Geology* 45 (5): 1502–1510.
19. Seethamraju, S., and S. Bandyopadhyay. 2023. Hydrogen economy: Hope or hype? *Clean Technologies and Environmental Policy* 25 (3): 755–756.
20. Liu, J. Y., et al. 2023. Genesis and energy significance of natural hydrogen. *Unconventional Resources* 3: 176–182.
21. Hren, R., et al. 2023. Hydrogen production, storage and transport for renewable energy and chemicals: An environmental footprint assessment. *Renewable and Sustainable Energy Reviews* 173: 113113.
22. Jin, Z., et al. 2024. Discovery of anomalous hydrogen leakage sites in the Sanshui Basin, South China. *Science Bulletin* 69 (9): 1217–1220.
23. Rigollet, C., and A. Prinzhofer. 2022. Natural hydrogen: A new source of carbon-free and renewable energy that can compete with hydrocarbons. *First Break* 40 (10): 78–84.
24. Martin, P. 2023. Natural hydrogen detected in "multiple locations" in South Australia. HydrogenInsight.
25. Blay-Roger, R., et al. 2024. Natural hydrogen in the energy transition: Fundamentals, promise, and enigmas. *Renewable and Sustainable Energy Reviews* 189: 113888.
26. Langhi, L., and J. 2023. Strand, exploring natural hydrogen hotspots: A review and soil-gas survey design for identifying seepage. Geoenergy 1.
27. Prinzhofer, A., et al. 2019. Natural hydrogen continuous emission from sedimentary basins: The example of a Brazilian H_2-emitting structure. *International Journal of Hydrogen Energy* 44 (12): 5676–5685.
28. Madadi Avargani, V., et al. 2022. A comprehensive review on hydrogen production and utilization in North America: Prospects and challenges. *Energy Conversion and Management* 269: 115927.
29. Roque, B. A. C., et al. 2025. Hydrogen-powered future: Catalyzing energy transition, industry decarbonization and sustainable economic development: A review. *Gondwana Research* 140: 159–180.

30. Prinzhofer, A., C. S. T. Cissé, and A. B. Diallo. 2018. Discovery of a large accumulation of natural hydrogen in Bourakebougou (Mali). *International Journal of Hydrogen Energy* 43 (42): 19315–19326.
31. Tilley, D. 2013. Geology for Investors. Gravity Surveys.
32. Pourtsidou, A. 2016. Testing gravity at large scales with H I intensity mapping. *Monthly Notices of the Royal Astronomical Society* 461 (2): 1457–1464.

2 Geological Environment

Natural hydrogen (white hydrogen) is found in diverse geological environments (Fig. 2.1), originating from distinct geological processes that drive its generation, migration, and entrapment [1]. Natural hydrogen accumulations have been identified in diverse geological settings, including igneous rocks, kimberlites, ore deposits, Precambrian basement units, salt structures, sedimentary and metamorphic sequences, ultramafic rocks, volcanic systems, and coal basins, each characterized by distinct geologic conditions that favor hydrogen generation, migration, or retention [2, 3]. The following sections provide a comprehensive analysis of these formation types, emphasizing their H_2 generating mechanisms, critical influencing factors, and illustrative case studies where applicable.

2.1 Geological Setting

Natural hydrogen systems are commonly linked to distinct geological environments where subsurface conditions are conducive to the production, transport, and storage of H_2 (Fig. 2.2 and Table 2.1) [4]. Notable examples of such reservoirs are found in diverse regions across the globe, including: (i) Ophiolite Complexes: Regions like the Oman Ophiolite Complex are globally recognized for hosting active serpentinization processes, a geochemical reaction in which H_2 is produced through the hydration of ultramafic rocks. These terrestrial environments are considered prime targets for hydrogen exploration due to the natural exposure of deeply-seated geological formations, which facilitates the study and extraction of naturally occurring H_2 [5]. (ii) Precambrian Shields: Geologically stable and ancient cratonic regions, such as the Canadian and Fennoscandian Shields, host abundant radioactive minerals, rendering them highly conducive to H_2 production

Y. Yuan et al., *Natural Hydrogen*, Synthesis Lectures on Renewable Energy Technologies, https://doi.org/10.1007/978-3-032-14469-0_2

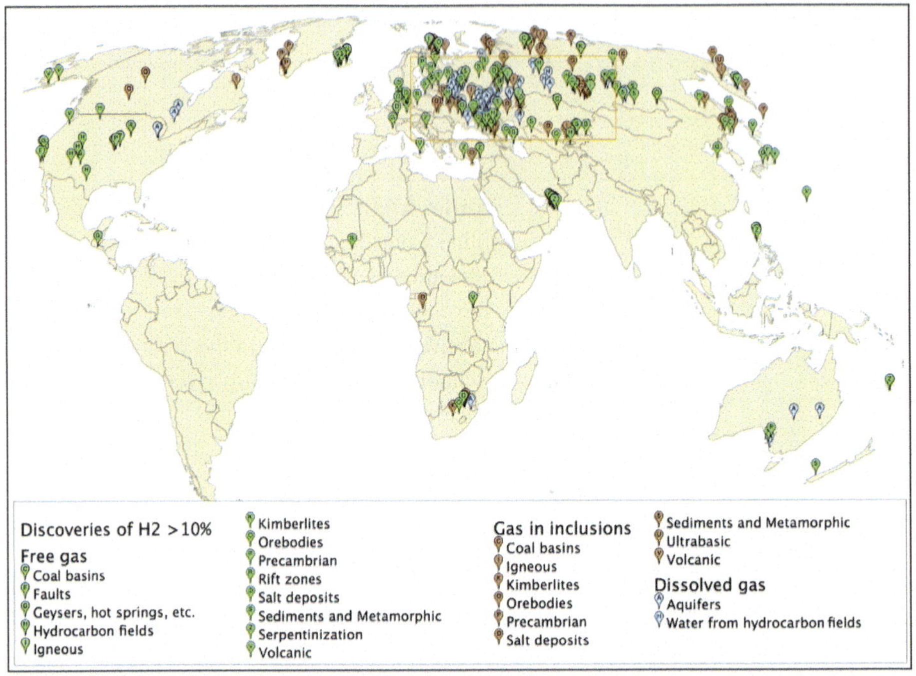

Fig. 2.1 The natural hydrogen system localities across the world. Geospatial distribution of hydrogen concentrations exceeding 10% vol detected in fluid inclusions across diverse environments (reproduced with permission from [1])

via radiolysis. These regions' enduring geological stability creates optimal conditions for prolonged hydrogen accumulation over geological timescales [6]. (iii) Volcanic and Geothermal Areas: Geothermally active regions (e.g., the East African Rift), are characterized by large mantle degassing associated with volcanism, which facilitates the release of molecular hydrogen (H_2). These areas are marked by dynamic tectonic systems that enhance the migration of H_2 [7]. (iv) Sedimentary Basins: Several sedimentary basins, including the Appalachian Basin in the USA and the Witwatersrand Basin in South Africa, are identified for microbially-mediated hydrogen generation. Their organic-rich shale deposits and carbonaceous layers foster anaerobic microbial processes that facilitate localized H_2 generation [8]. (v) Salt Domes and Evaporites: Salt deposits serve as exceptional geological barriers for hydrogen entrapment, though natural hydrogen production predominantly originates in neighboring geological strata or deeper subsurface settings. Multiple geochemical processes drive H_2 formation, with subsequent migration into salt-associated reservoirs occurring along fracture networks or fault systems. Notable examples such as the Gulf of Mexico and Europe's Zechstein Basin illustrate how salt formations function as highly efficient caprocks for H_2 containment. These evaporite

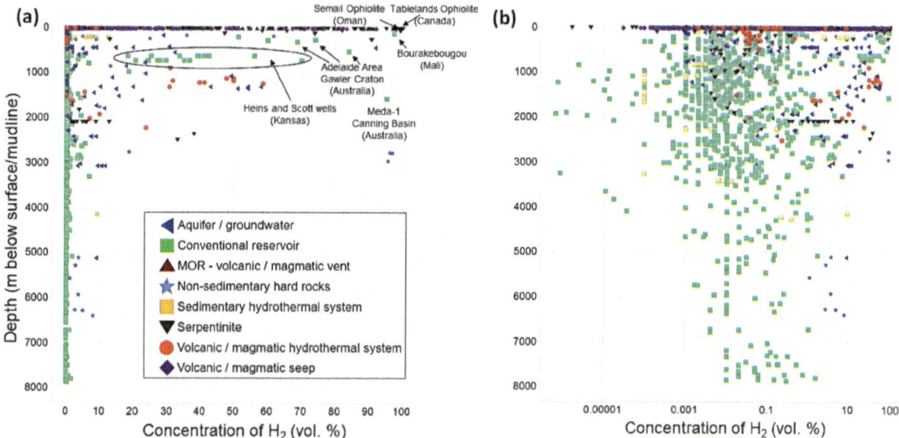

Fig. 2.2 Concentrations of H_2 are measured at the surface and at depth across diverse geological environments. **a** data presented on a linear scale. **b** Data illustrated using a logarithmic scale (reproduced with permission from [28])

structures exhibit extremely low permeability, effectively retaining H_2 that ascends from depth and maintaining long-term reservoir stability [9].

2.2 Occurrence State of Hydrogen

In the current exploration and exploitation of natural hydrogen, free hydrogen gas in surface environments has drawn extensive attention and serves as the primary utilization pathway. Additionally, the inclusion of hydrogen, dissolved hydrogen, and adsorbed hydrogen also demonstrates significant development potential [29].

2.2.1 Free Hydrogen

As a directly exploitable resource, free hydrogen gas constitutes the core research focus in contemporary studies of natural hydrogen. Surface manifestations of free hydrogen leakage typically present as elliptical depressions devoid of vegetation cover, commonly referred to as "fairy circles" [2]. These depressions likely result from H_2 migration through deep-seated pathways: the rock alteration along H_2 migration channels induces surface subsidence, forming elliptical depressions. During H_2 exploration in Mali, Africa, "fairy circles" were identified near high-concentration H_2 wells, with surrounding H_2 levels ranging from 0.001–0.6‰ [2]. Satellite imagery from Pingrupyi in Western Australia, its southern regions, and areas in South Australia similarly reveals characteristic

Table 2.1 Examples of natural hydrogen, with H_2 content exceeding 40% (adapted from [10])

Construct background	Hydrogen content (%)	Region/Country	Geological environment and formation mechanism	References
Stabilization zone	91.8	Kansas, United States of America	Precambrian rocks (deep-source hydrogen gas: water-reduction reaction associated with Fe oxidation; reactions related to high-content reduced iron within casing pipes)	[11]
	68.6	Penneshaw, Australia	Precambrian rocks (–)	[12]
	57.8	Sudbury, Canada	Precambrian rocks (radiolytic and hydration reactions)	[13, 14]
	98	Bourakebougou, Mali	Sedimentary rocks (inorganic genesis related to Neoproterozoic sedimentary rocks)	[2]
	75.8	Poison Bay, New Zealand	Sedimentary rock (serpentinization reaction)	[15]
	43.8	Chuxiong Basin, China	Sedimentary rocks (–)	[16]
Construction activity area	50.9	Camp Spring, America	Serpentinite (serpentinization of shallow and deep source water)	[17]
	97	Bahla, Oman	Serpentinite (gas leakage from surface serpentinization)	[18]
	48.3	Vaiceva Voda, Bosnia and Herzegovina	Serpentinite (serpentinization of rock without water or containing unsaturated metallic catalysts)	[19]
	96.3	Hoffman, United States of America	Rift Valley Region (inorganic reactions of Fe^{2+})	[20]
	80.4	Iriklinskoe, Russia	Igneous rock (–)	[1]
	80.5	Nizhny Tagil, Russia	Igneous Rock (-)	[21]
	57.3	Namafjall, Iceland	Rift Valley Region (reaction of reducing carbon and water in magma)	[22]

(continued)

Table 2.1 (continued)

Construct background	Hydrogen content (%)	Region/Country	Geological environment and formation mechanism	References
	57.8	Etna, Italy	Volcanic gases (vent)	[23]
	51.5	Augustine, United States of America	Volcanic gases (shallow crustal sedimentary rocks)	[24]
	51.4	Arima, Japan	Fountains and Hot Springs (–)	[25]
	100	Cyprus	Ore body (pod-like chromite in the mantle)	[26]
	100	Oklo, Gabon	Mineral body (–)	[27]
	61.5	Muhlhausen, Germany	Salt mine (mixed origin)	[21]
	>50	Wittlwheim, France	Salt Mine (–)	[1]

"fairy circle" features [30]. Soil gas investigations in North Carolina, United States, have also detected "fairy circles" associated with substantial free hydrogen emissions [31]. The global distribution of these "fairy circles" frequently corresponds with natural hydrogen system exhibiting elevated H_2 concentrations [29].

2.2.2 Inclusions Hydrogen

The gas composition in different geological bodies can indicate distinct geological origins [32, 33]. However, even within the same geological environment, H_2 concentrations exhibit significant variations [1]. The presence of H_2 in fluid inclusions is more readily observed in Precambrian formations, magmatic rocks, volcanic rocks, and various ore bodies, while fewer cases have been documented in sedimentary and metamorphic rocks [1, 34]. High-concentration H_2 inclusions have been identified in multiple geological settings [35]. H_2 content in ophiolite inclusions reaches 15–100%, with maximum rock H_2 concentrations up to 6.3 cm^3/kg. Precambrian rock inclusions may contain over 90% H_2, while whole-rock H_2 concentrations can exceed 40 cm^3/kg. However, challenges remain in determining whether this inclusion-hosted H_2 could serve as directly exploitable natural hydrogen due to limited bulk-rock H_2 content and restricted distribution of fluid inclusion-bearing rocks. This necessitates the identification of H_2 reservoirs sourced from inclusion H_2 and evaluation of their potential for economic exploitation.

2.2.3 Dissolved Hydrogen

Research on dissolved hydrogen primarily focuses on its occurrence in underground aquifers, oil and gas field waters, hot springs, and geothermal wells. Although the relationships between dissolved hydrogen concentrations and parameters such as depth, salinity, pH values, rock types, borehole age, fractured aquifer age, or other measurements remain unclear, current studies indicate that the highest concentrations are typically found in deeper, hypersaline, and older waters within fractured aquifers [36, 37]. Previous investigations have demonstrated that groundwater near oil fields contains higher dissolved hydrogen content compared to that near gas fields [1]. This occurrence could be linked to the enhanced hydrogen-producing capacity of organic matter within oil-bearing formations [38], thereby providing more abundant H_2 sources for oil-field waters. Substantial quantities of free hydrogen gas have also been detected within hydrothermal springs and geothermal reservoirs [39]. Analytical measurements of these gaseous samples reveal substantial free hydrogen content, suggesting indirectly that H_2 exhibits relatively high solubility in high-temperature aqueous environments [29]. However, investigations have revealed that dissolved hydrogen in groundwater can serve as a buffer layer for temporary H_2 storage [40]. This layer temporarily entraps H_2 derived from hydrogen-generating source rocks and other origins, while simultaneously replenishing free-phase H_2 accumulations in shallow geological formations within H_2 systems [41]. The existence of such a buffer layer proves critical for maintaining dynamic H_2 systems, primarily due to hydrogen's inherent physical properties (notably low density and strong reducibility) [42]. Through its capacity to cyclically sequester and release H_2 from multiple sources, this natural buffering mechanism facilitates sustainable H_2 migration processes, thereby enabling the feasible extraction and utilization of natural hydrogen system.

2.2.4 Adsorbed Hydrogen

H_2 primarily exists in an adsorbed state within mesopores and micropores, with its occurrence characteristics being closely associated with lithology, pore structure, temperature, and pressure conditions [43]. Clay minerals, which are diverse in type and widely distributed underground, may exhibit considerable potential for H_2 adsorption due to their large specific surface areas [29].

H_2 adsorption efficiency in coal-bearing rocks is potentially enhanced by their abundant organic matter content and well-developed micropores. The high pore connectivity of coal matrices facilitates H_2 ingress and subsequent adsorption. During the thermal maturation of organic matter, substantial H_2 radicals are liberated, leading to copious H_2 production as associated gas with hydrocarbons [44]. Furthermore, H_2 supplementation can augment hydrocarbon generation efficiency from kerogen, particularly at high-overmature stages, where exogenous H_2 input significantly increases alkane gas yields [45]. Experimental studies demonstrate positive correlations between H_2 adsorption capacity and three key parameters: thermal maturity level, organic matter abundance, and micropore development intensity [38]. Notably, thermal maturity and organic richness fundamentally govern micropore architecture and specific surface area characteristics in coal deposits [29].

Carbonate rocks, widely distributed globally, are prone to karst processes that develop distinct porosity and permeability characteristics. Exploration in Mali has revealed that dolomitic carbonate rocks form the shallowest principal H_2 reservoirs, with substantial free-state H_2 preserved in their karstic pore systems [3]. Additionally, the presence of water films on carbonate surfaces further inhibits H_2 adsorption. Consequently, carbonate formations exhibit relatively low adsorbed hydrogen content, which shows significant sensitivity to temperature, pressure, competitive adsorption, and aqueous film effects [29].

Overall, in various rock types, increased pressure generally enhances H_2 adsorption capacity, whereas elevated temperatures tend to reduce it [46]. Although the accumulation capacity of adsorbed hydrogen is relatively limited and insufficient to serve as a primary reservoir for H_2 extraction, it nevertheless functions similarly to dissolved H_2 as an effective buffer system. This adsorbed hydrogen can concentrate and integrate H_2 sourced from diverse geological origins. Significantly, when subjected to alterations in temperature–pressure conditions or geological events, this adsorbed hydrogen can be mobilized and released through conversion into free-state H_2.

2.3 Global Distribution of Hydrogen Discoveries

Hydrogen has traditionally been considered rare in its free gaseous form in natural subsurface settings. However, it can be generated through various geological processes, and historical evidence from drilling operations and natural seeps indicates occurrences of elevated hydrogen fluxes. Nevertheless, conventional gas sampling and analysis have seldom targeted hydrogen, leading to its underreporting and resulting in a systematic bias in existing databases (Fig. 2.3).

2.3.1 Major Natural Hydrogen Areas and Exploration Potential Zones

Given that natural hydrogen reservoirs are typically associated with their geological settings, the global distribution of such reservoirs and regions with exploration potential can be classified based on variations in hydrogen sources (Fig. 2.4) and seepage characteristics (Fig. 2.5) across different environments (Table 2.2). This enables a comprehensive assessment of the sustainability of hydrogen energy and its prospects for resource development.

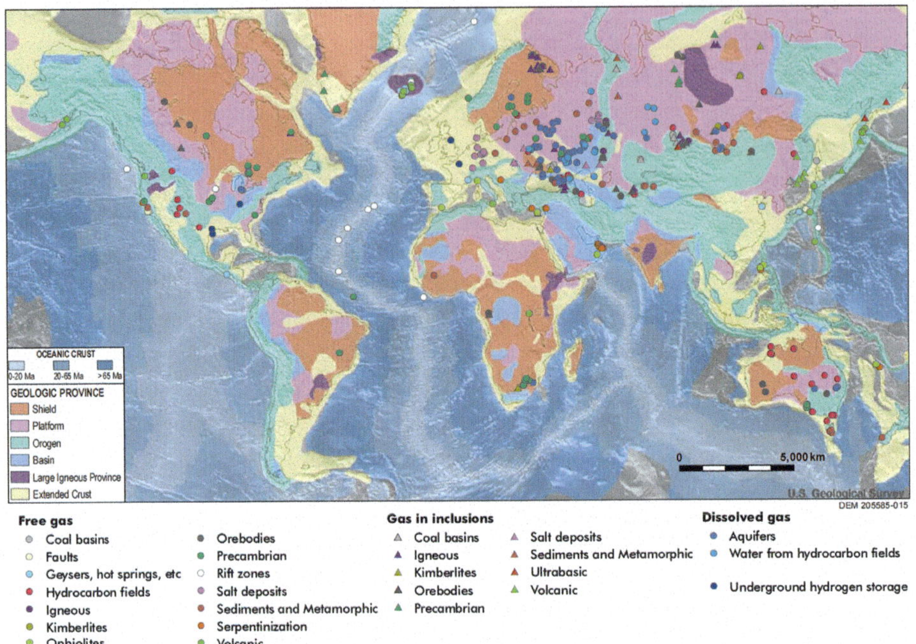

Fig. 2.3 Locations and geological settings worldwide where hydrogen concentrations exceeding 10% by volume have been documented. As highlighted by Zgonnik, the clustered distribution of H_2 occurrences across Europe and Asia does not necessarily indicate higher innate prospectivity for molecular hydrogen, but rather reflects a sampling bias in existing data collection efforts [47]. (Modified from [1, 13, 30, 48–51])

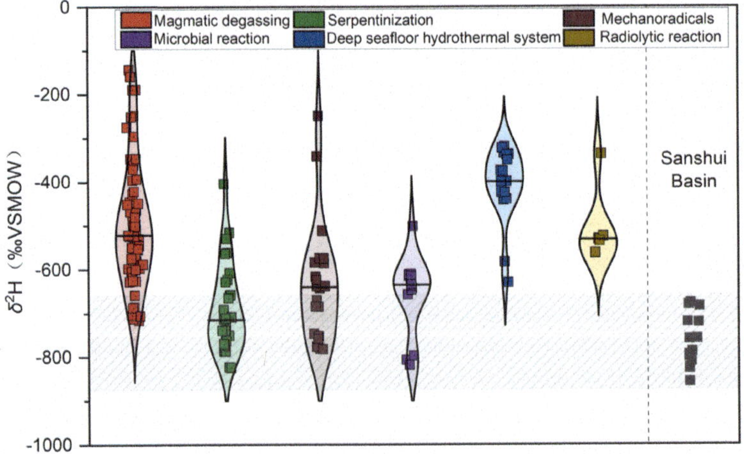

Fig. 2.4 δ^2H values of H_2 with different origins [52, 53]

Fig. 2.5 Hydrogen emission under different environments. **a** The perennial gas seep at Chimaera, located within the Antalya region of Turkey, exhibits sustained combustion [1]. **b**, **c** Submarine fumaroles and terrestrial fumaroles for hydrogen [54]. **d** Hydrogen phenomenon at hot spring entrance [54]

2.3.1.1 Mid-Ocean Ridges and Oceanic Crust

The natural hydrogen system generated at mid-ocean ridges and within oceanic crust are primarily formed through serpentinization, a geological process in which ultramafic rocks undergo hydration via reaction with seawater. Mid-ocean ridges, including the Mid-Atlantic Ridge and the East Pacific Rise, are tectonically dynamic settings marked by the ongoing production of new oceanic crust [77]. Within high-temperature, high-pressure environments, the aqueous alteration of olivine-rich peridotites yields serpentine minerals while liberating molecular H_2 [55]. The generated H_2 subsequently accumulates in crustal fractures and pore networks, establishing potential natural reservoirs. These systems represent optimal sites for sustained H_2 production, attributed to persistent geothermal gradients coupled with active geological processes [78].

Hydrothermal vents are fissures on the seafloor, mainly distributed along mid-ocean ridges and other tectonically active zones, that emit high-temperature fluids of geothermal origin. These systems are critical for the formation of natural hydrogen reservoirs via the process of serpentinization [79]. Under these settings, seawater circulates through

Table 2.2 The seven major distribution areas and exploration potential zones of natural hydrogen gas reserves

Geological environment		Geological Background	Instances and causes have been identified	Example
Constructing activity area	Mid-ocean Ridges	Mid-ocean ridges represent the primary loci of seafloor spreading, where the continuous accretion of new oceanic crust occurs through magmatic. processes Ultramafic rocks (pyroxene, olivine) interact with seawater and undergo serpentinization, producing substantial amounts of H_2	The East Pacific Rise at 21° North; the Atlantic Mid-Atlantic Ridge rainbow, attributed to serpentinization [55]; the Southwest Indian Ridge [56]; Lost City, attributed to serpentinization [57]	The Lost City hydrothermal field, situated in the southern part of the Azores, has revealed significant natural H_2 reserves. This area is primarily characterized by exposed ultramafic rocks that have undergone intense serpentinization, creating essential conditions for H_2 generation. Measurements indicate exceptionally high H_2 concentrations in vent fluids from the Lost City hydrothermal system, typically ranging from 10 to 15 mmol/kg, with peak values reaching up to 40 mmol/kg

(continued)

Table 2.2 (continued)

Geological environment		Geological Background	Instances and causes have been identified	Example
	Sector convergence zone	The oceanic plate is subducting beneath the continental plate, leading to intense geological activity (volcanic eruptions and magma movement). These geological processes can facilitate the generation of H_2	Chimaera region, Turkey, serpentinization [58]; New-Caledonia, Australia; Zambales, Philippines, serpentinization [59]; Omani ophiolite belts (e.g., Bahla and Huwayl Qufays), serpentinization [60]	The Chilean convergent margin is defined by the subduction of the Nazca Plate beneath the South American Plate, characterized by intense tectonic activity accompanied by frequent earthquakes and volcanic eruptions. In recent years, significant natural hydrogen concentrations have been discovered in specific deep-seated faults and serpentinization zones within the Chilean Subduction Zone, with primary detection sites located at the Atacama Trench and the Central Chile Fault Systems

(continued)

Table 2.2 (continued)

Geological environment		Geological Background	Instances and causes have been identified	Example
	Fault zone	A fault zone is an area where rocks are fractured and displaced. Fault zones usually connect the deep parts of the Earth's crust to the surface, forming good gas migration pathways. H_2 produced in the deep mantle and H_2 produced from serpentinization can be transmitted to the surface	Near the fault zones of the Jiyang Depression in the Bohai Bay Basin, the genesis is partly attributed to mantle-derived degassing [45]; the Xujiaweizi fault depression [61]; the eastern Tanlu fault zone [62]; the Yilan-Yitong fault [63]	\
	Continental rift	The continental rift is a geological structure formed by the stretching and fracturing of the continental crust under tensile forces. The tension in the crust exposes deep rocks to the action of water, potentially resulting in the generation of H_2. Additionally, the rift provides pathways for the escape and accumulation of H_2	Kansas Island in North America; Willey 1 well and Hofmann 3 well located approximately 100 km northwest of Iowa, USA [64]; Perth Basin; San Shui Basin; the Kansas segment of the USA Mid-Continent Rift, formed due to serpentinization and mantle degassing [20]; Hengill area in Iceland, formed due to serpentinization [20]	The East African Rift Valley is one of the most active and typical continental rift systems in the world. Substantial natural hydrogen anomalies have been detected across multiple sectors of the rift system over recent years. In certain hot springs and vents, the concentration of H_2 can reach hundreds of ppm or even higher. These high concentrations of H_2 typically occur near the active tectonic zones and volcanic centers of the rift

(continued)

Table 2.2 (continued)

Geological environment		Geological Background	Instances and causes have been identified	Example
	Magma related	The high-temperature and high-pressure conditions associated with magmatic processes, water–rock interactions, and deep Earth degassing may collectively contribute to the generation and release of H_2	Lanier Devonian eruptive rocks; 11.7% H_2 was detected in volcanic gas from Sandorina, Italy [65]; volcanic rocks in northern Oman [66]; Khibina and Lovozero, Greenland, Kola Peninsula, northwestern Russia Alkaline intrusive complexes such as Ilímaussaq and Strange Lake, Quebec, Canada [67]	\
	Hot spring	Serpentinization, thermal decomposition of water under high-temperature conditions, and mantle degassing all provide abundant H_2. Meanwhile, geothermal areas are often accompanied by tectonic activity, and the fractures created by tectonics provide pathways for the ascent of H_2	Geothermal gas in Tengchong Hot Sea, China [68]; geothermal gas in Tuoba Town, Garze, Sichuan [69]; Tianchi in Changbai Mountain [70]; Wudalianchi in Heilongjiang	In multiple hot springs and thermal springs in Tengchong, Yunnan, Southwest China, significant concentrations of H_2 have been detected through gas sampling and analysis. The discovery of this H_2 indicates that Tengchong's geothermal system not only produces traditional geothermal gases but may also be an important generation area for natural hydrogen

(continued)

Table 2.2 (continued)

Geological environment		Geological Background	Instances and causes have been identified	Example
Constructing activity area	Precambrian shield	The formation of H_2 in the Precambrian cratons is primarily achieved through various deep geological and chemical processes, which include fluid-rock interactions, radioactive decay, and tectonic movements	The Canadian Shield Kidd Creek mining area; The Precambrian basement rocks underlying the Witwatersrand Basin in South Africa; the Precambrian shield of Scandinavia in Finland; the Precambrian shield in Kansas, USA; and Yorke Peninsula in South Australia, caused by hydrothermal alteration and radiogenic decay [71]	The Kidd Creek mining area in the Canadian Shield has gases in groundwater at depths of 2072 to 2100 m below the surface, containing 0.40% to 12.7% H_2; in the Witwatersrand basin of South Africa, the dissolved gases in groundwater from Precambrian shield rocks contain H_2, with a maximum proportion of up to 11.5% in the gas composition
	Sedimentary basins	The prolonged subsidence of ancient stable cratonic regions has resulted in significant sediment accumulation. The generation of natural hydrogen within these basins may be associated with deep crustal fractures and fissures, H_2 production from organic matter, as well as water–rock interactions between groundwater and basement lithologies. The stable tectonic setting of these basins facilitates effective H_2 preservation and accumulation	Bourakebougou H_2 field in Mali, West Africa, is characterized by serpentinization of base diabase [72]; Craton Basin, southeastern Brazil [72]; Carolina Bay, North Carolina, USA [31]; Songliao Basin in China, the result of a mixture of multiple genesis [73]; Well No. 2 in the Sanhu area of the Qaidam Basin, China [74]; Chuxiong Basin in China [16]; the New Guinea region of Australia [75]; Stavropol in southern Russia [76]	\

the oceanic crust and interacts with ultramafic rocks, such as peridotite. The elevated thermostatic temperatures and lithostatic pressures promote reactions between water and olivine-rich lithologies, thereby driving serpentinization and the generation of H_2 [55].

Within the oceanic crust, H_2 accumulates in fractures and porous spaces, forming potential reservoirs. Persistent geothermal activity, coupled with dynamic geological conditions at hydrothermal vents, maintains a steady H_2 production rate [28]. A detailed discussion of processes such as serpentinization will be provided in the following sections.

Hydrothermal vents create distinct and extreme ecosystems where steep chemical gradients facilitate H_2 production. Metal-rich minerals within vent fluids act as natural catalysts, significantly enhancing H_2 synthesis [80]. These naturally occurring H_2 reservoirs represent a substantial reservoir for sustainable energy production, providing a source of clean, renewable energy that can be extracted with virtually no ecological disruption [28]. A comprehensive understanding of these geological systems and their exploration is essential for unlocking this viable energy resource.

2.3.1.2 Subduction Zones

These dynamic geological environments result from the convergent boundaries of tectonic plates. In these tectonic regimes, subduction zones develop as an oceanic plate descends beneath either a continental plate or an adjacent oceanic plate, generating extreme conditions characterized by elevated pressures, temperatures, and fluid-mediated processes [48]. During slab descent, the subducting plate transports aqueous fluids and volatiles into the mantle. The progressive increase in pressure and temperature triggers dehydration reactions within the descending slab, liberating aqueous fluids enriched with dissolved minerals and chemical species [81]. These metasomatic fluids ascend through the lithosphere, interacting with the ultramafic lithologies (e.g., peridotite) of the overlying mantle wedge. These fluid–rock exchanges drive serpentinization reactions, a process critical to mantle geochemistry and ore formation [82]. For example, the regions of Oman, Luzon in the Philippines, New Caledonia, and Turkey are illustrated in the Fig. 2.6.

2.3.1.3 Cratonic Regions (Stable Continental Interiors)

The Kla Peninsula in northwestern Russia hosts ancient Precambrian shields, where natural hydrogen seeps have been detected along deep-rooted fractures and fault systems. Similarly, the Volga-Ural Region (Russia) and Witwatersrand Basin (South Africa) represent cratonic regions distinguished by their ancient and stable geological structures, which are increasingly recognized for their potential to host natural hydrogen system. Three prominent cratonic areas, the Kola Peninsula (Russia), Volga-Ural Region (Russia), and Witwatersrand Basin (South Africa), have emerged as key targets for H_2 exploration due to their unique geological histories and structural frameworks [1]. (i) Kola Peninsula: The Kola Peninsula exhibits remarkable geological diversity and a widespread occurrence of Precambrian rock formations. Research has documented substantial H_2 degassing in this region, particularly correlated with deep-seated fractures and fault systems. The interplay

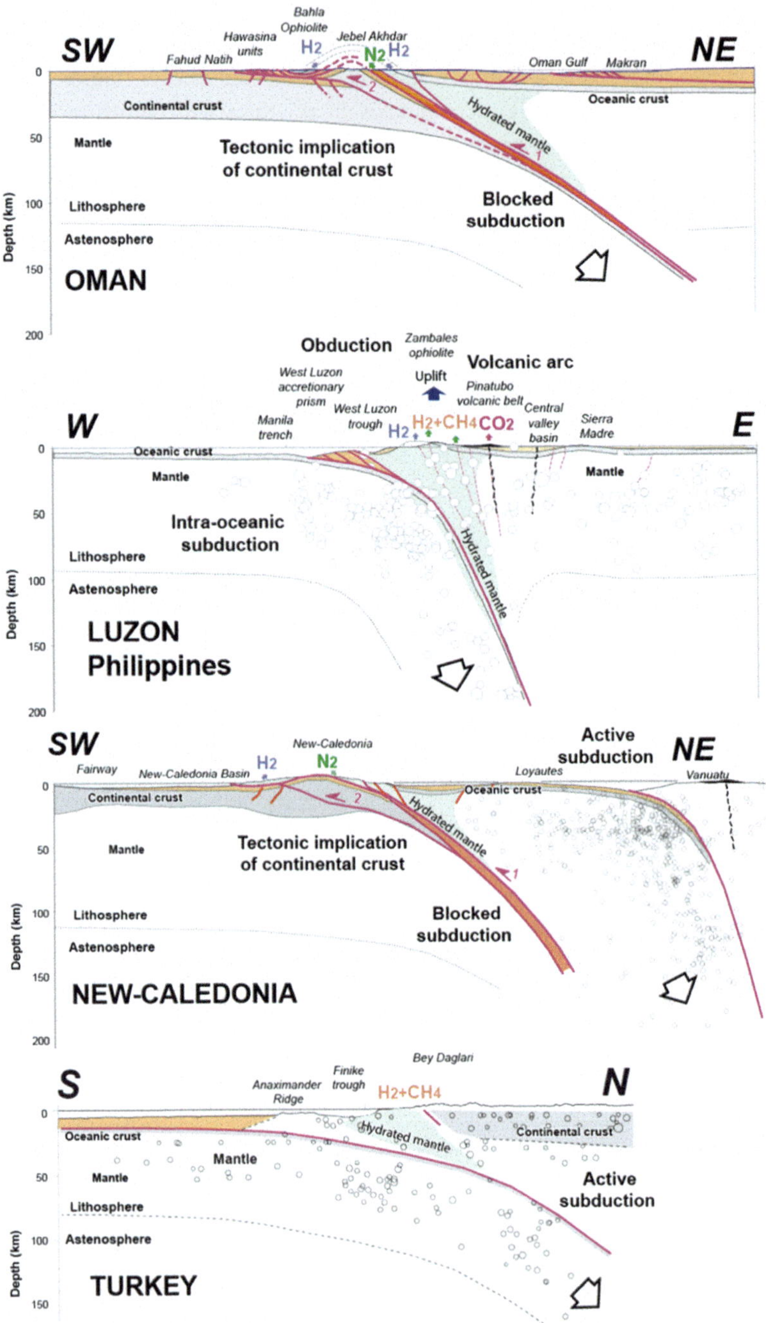

Fig. 2.6 Schematic diagram of hydrogen generation in subduction zones in Oman, Luzon, Philippines, Nova Caledonia, and Turkey [83]

between the area's stable cratonic environment and active geothermal gradients establishes favorable conditions for natural hydrogen generation, migration, and accumulation [84]. (ii) Volga-Ural Region: Located in the southwestern part of Russia, the Volga-Ural Region constitutes a segment of the East European Craton. This area is characterized by extensive sedimentary basins and substantial hydrocarbon reserves. The presence of ancient crystalline basement rocks, coupled with active tectonic processes, provides a favorable geological setting for the generation of natural hydrogen. Recent research studies indicate that natural hydrogen is frequently linked to conventional oil and gas reservoirs, implying opportunities for co-production during hydrocarbon extraction [85]. (iii) Witwatersrand Basin: Renowned for its abundant gold reserves and well-preserved Precambrian lithologies, the Witwatersrand Basin in South Africa represents a promising target for natural hydrogen exploration, owing to its long-term geological stability and complex tectonic evolution. Subsurface H_2 occurrences have been identified across the basin, exhibiting spatial correlations with crustal-scale structural discontinuities such as fault systems and fracture networks [86].

In comparison, all three regions exhibit key similarities, including ancient crustal structures, tectonic quiescence, and prominent fault systems, which play a vital role in generating and concentrating natural hydrogen [1]. Nonetheless, their divergent geological evolution and site-specific environmental factors modulate H_2 occurrence and resource accessibility across these regions. Recognizing these distinctions is essential for designing tailored exploration approaches and development frameworks to maximize the potential of natural hydrogen reserves in cratonic settings.

2.3.1.4 Sedimentary Basins

In sedimentary basins, H_2 generation occurs through multiple key geological mechanisms, as detailed in the subsequent chapter. Fundamentally, this process is primarily driven by the thermal maturation of organic matter, whereby organic-rich lithologies such as shales and coals undergo decomposition under high-temperature and high-pressure conditions, liberating H_2 [87].

H_2 generation in subsurface environments is driven by multiple interconnected processes. Water–rock interactions are critical mechanisms, particularly through reactions between water and ferrous iron-bearing minerals (e.g., olivine, pyroxene), which liberate H_2 [88]. Radiolysis, the radiation-induced dissociation of water molecules through radioactive decay of isotopes such as uranium, thorium, and potassium, further supplements H_2 production [89]. Microbial activity in subsurface anaerobic environments facilitates this mechanism, whereby certain microorganisms produce H_2 as a metabolic by-product [87]. These processes operate in concert to enable the migration and subsequent accumulation of molecular hydrogen through porous and permeable sedimentary reservoirs. Structural traps (e.g., anticlines, synclines) and impermeable caprocks act as natural seals, enabling reservoir formation [88]. Notable examples include the Hugoton-Panhandle gas field in Kansas, United States of America, where H_2 was documented during natural gas extraction [87], and the São Francisco Basin in Brazil, where H_2 generation is hypothesized to result from deep-seated geological processes, including organic matter decomposition and water–rock interactions driving serpentinization reactions.

2.3.1.5 Volcanic Regions

H_2 generation in volcanic regions is driven by multiple interconnected processes. A dominant mechanism involves chemical exchanges between volcanic gases and groundwater systems. When magma interacts with surface or subsurface water reservoirs, hydrothermal reactions are initiated, leading to H_2 synthesis [1]. Another critical contributor is serpentinization, a reaction in which ultramafic rocks (e.g., olivine-rich formations) undergo hydration under elevated temperatures and pressures, liberating molecular H_2 [90]. Volcanic zones also exhibit enhanced geothermal gradients, which promote water molecule dissociation via radiolysis, a process powered by radioactive decay processes within volcanic lithologies. Direct mantle-derived H_2 release further supplements these systems, occurring during eruptions or through fumarolic degassing, where H_2 escapes alongside other mantle volatiles. These combined mechanisms enable H_2 migration and sequestration in permeable volcanic rock matrices and fracture networks, culminating in natural hydrogen system [91]. This phenomenon is exemplified by the Izu Bonin Mariana Arc System in Japan, where geothermal fields exhibit substantial H_2 emissions from volcanic activity. Similarly, the Hengil Volcanic Zone in southwestern Iceland serves as a modern analog, hosting active H_2 generation linked to geothermal–volcanic interactions.

2.3.1.6 Ophiolites or Fragments of Oceanic Crust on Land

Natural hydrogen in ophiolitic formations, such as the Semail Ophiolite in Oman and the Troodos Ophiolite in Cyprus, are generated via the geochemical process of serpentinization. This process entails hydration reactions occurring between ultramafic rocks, which are rich in ferromagnesian minerals like olivine and pyroxene, and aqueous fluids [5, 92]. When these lithologies interact with aqueous solutions under particular physicochemical conditions, such as the elevated temperature and pressure regimes characteristic of ophiolite systems, they undergo mineralogical transformation through serpentinization, generating serpentine group minerals and molecular H_2 [90].

Furthermore, tectonic activity associated with ophiolite emplacement creates networks of fractures and fault systems that provide primary pathways for the migration and subsequent accumulation of hydrogen within geological reservoirs [92]. This synergy of geochemical reactions and structural conditions establishes ophiolites as optimal settings for the production and sequestration of natural hydrogen. The Semail and Troodos Ophiolites, characterized by vast ultramafic rock outcrops and active tectonic regimes, represent prominent examples where such mechanisms result in the development of substantial natural hydrogen [93].

2.3.1.7 African Rift Zones

The Rungwe Volcanic Province in Tanzania and the Bourakebougou area in Mali serve as prominent examples of natural hydrogen system, showcasing the varied geological mechanisms driving H_2 generation and accumulation. Situated within the East African Rift System, the Rungwe Volcanic Province exhibits significant volcanic activity and anomalously high geothermal gradients. H_2 formation here is predominantly attributed to hydrothermal reactions and serpentinization processes [94, 95]. Volcanic emissions promote interactions between magmatic gases and groundwater, accelerating H_2 synthesis. Furthermore, aqueous fluids interact with olivine- and pyroxene-bearing volcanic rocks in the region under elevated temperatures and pressures, leading to serpentinization, which generates serpentine group minerals and molecular hydrogen. The East African Rift's active tectonics generate extensive fracture and fault networks, which act as conduits for H_2 migration and storage, solidifying the Rungwe Volcanic Province as a critical hub for natural hydrogen system [95].

Conversely, the Bourakebougou region in Mali, situated within the West African Craton, produces H_2 via the interplay of radiolysis and geodynamic phenomena occurring in Archean crystalline basement rocks [94]. The Archean basement in this area hosts significant concentrations of radiogenic elements, including uranium, thorium, and potassium. The radioactive decay of these isotopes generates ionizing radiation, which induces water molecule dissociation into H_2 and O_2. Geodynamic processes in the region, such as recurrent tectonic activity, have generated extensive fracture networks and fault systems, providing conduits for H_2 migration from deep-seated reservoirs to shallow crustal levels. Field studies in Bourakebougou have documented prolific H_2 seepages along structural discontinuities and fracture zones within the Archean basement. This naturally occurring H_2 is harnessed through subsurface extraction systems and employed locally for electricity production, supplying renewable electricity to nearby settlements [94]. Bourakebougou has emerged as an exemplar of decentralized, community-driven H_2 exploitation, revealing the prospects of natural hydrogen as an emissions-free energy resource [96].

The Rungwe Volcanic Province and Bourakebougou area exemplify the diversity of geological settings and mechanisms that generate natural hydrogen system. In the Rungwe Volcanic Province, H_2 formation is linked to active volcanic and hydrothermal processes, whereas in Bourakebougou, it originates from radiolysis within ancient crystalline basement rocks [94]. Despite contrasting geological drivers, both regions underscore the critical role of tectonic activity in facilitating H_2 migration and reservoir development through fracture networks. Investigating the geological controls and dynamics of natural hydrogen systems is essential to advancing H_2 as a sustainable energy resource, which could play a pivotal role in reducing reliance on fossil fuels and supporting global decarbonization goals.

2.3.2 The Geological Environment of Natural Hydrogen Discovery

H_2 exhibits properties such as high volatility, low density, and strong reducibility, which make it extremely difficult for natural hydrogen to exist in a stable free state. Consequently, conventional perspectives held that natural hydrogen could hardly be preserved at Earth's surface. Simultaneously, the nascent stage of H_2 detection technology has hindered effective identification and localization of H_2 reservoirs, impeding the development of natural hydrogen exploration [97]. However, previous studies and explorations have demonstrated widespread natural hydrogen occurrences across the globe (Fig. 2.7). For instance, H_2 flow was detected in hydrothermal vents (commonly known as "black smokers" due to sulfide-rich fluids forming dark particulate precipitates upon cooling) erupting from the Mid-Atlantic Ridge's plate spreading zone in 1977 [98]. This H_2 generation is closely associated with serpentinization processes and reactions between water and sulfide minerals. The aforementioned Mali H_2 Field, successfully applied in civil applications, was recognized as one of Science's Top 10 Scientific Breakthroughs of 2023. Furthermore, the formation of "fairy circles" (circular surface depressions) discovered worldwide has been partially attributed to H_2 activity by some researchers (as illustrated below). H_2 flows have been detected in regions including North Carolina (USA), Perth Basin (Western Australia), Central Russia, and Southern Gironde (France) (Fig. 2.8a–c), with maximum H_2 concentrations reaching 1000 ppm [99]. However, dissenting scholars argue that the H_2 hypothesis represents merely one explanation for fairy circle formation, competing with alternative theories such as termite activity. At the same time, not all high-hydrogen areas will have corresponding circular depression characteristics on the surface, exemplified by the Meda-1 well in the Canning Basin of Western Australia, which reached a depth of 2685 m and recorded 95% H_2, and the Mt Kitty 1 well in the Amadeus Basin of the Northern Territory, drilled to 2295 m with 11% H_2, these occurrences provide clear evidence of natural hydrogen accumulation (Fig. 2.8d, h) is a good demonstration [100]. The aforementioned circumstances indicate that natural hydrogen system exhibit characteristics such as wide distribution and significant variations in concentration. High-concentration natural hydrogen is commonly found in the following geological environments (Fig. 2.9) [64]. Based on differences in tectonic settings, natural hydrogen distribution areas can be classified into two major categories: Tectonically active zones (e.g., mid-ocean ridges, plate convergent zones (ophiolite belts), magmatic and hydrothermal activity areas, and geological fractures and faults), and tectonically stable zones (e.g., Precambrian continental basements and sedimentary basins) [67].

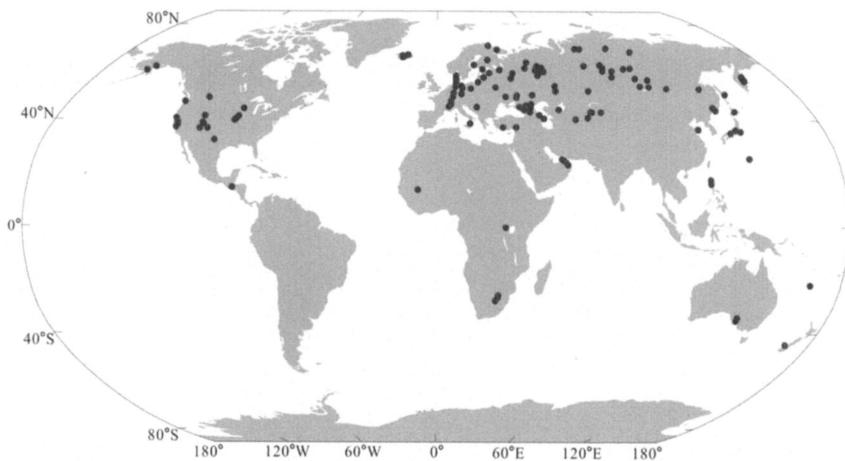

Fig. 2.7 Distribution of known hydrogen reservoirs with a volume fraction greater than 10% [1, 101]

2.4 Kimberlites Environment

Kimberlite is an ultramafic volcanic rocks that originates deep in Earth's mantle and is commonly found in ancient continental cratons [110]. Narrow vertical pipes (known as kimberlite pipes) can often be formed, which can create potential migration pathways for H_2. H_2 in kimberlite environments is mainly derived from the serpentinization of water and mafic minerals in kimberlite, mantle-derived H_2 carried by kimberlite as it rises from the mantle to the surface, volcanic gases released during the formation of kimberlite pipes, and H_2 that migrates up along kimberlite pipes and accumulates in overlying traps. Its content is affected by the composition of the mantle, the depth of origin, and the content of volatiles [7].

Current studies highlight H_2 potential in cratonic regions like the Kaapvaal and Siberian Cratons, where H_2 seeps and mantle-derived volatiles have been detected near kimberlite fields. Geophysical techniques (e.g., seismic surveys) and geochemical analyses of fluid inclusions are advancing exploration. However, challenges persist, including limited data on H_2 generation mechanisms, complex pipe structures hindering reservoir characterization, and hydrogen's high mobility leading to leakage risks. Existing exploration technologies, optimized for hydrocarbons, require adaptation to detect hydrogen-specific geochemical and geophysical signatures. Additionally, identifying effective traps, such as impermeable caprocks, remains critical for viable extraction [111].

Despite the challenges, the kimberlite environment offers a promising pathway for natural hydrogen development. Growing interest in the craton region, coupled with advances

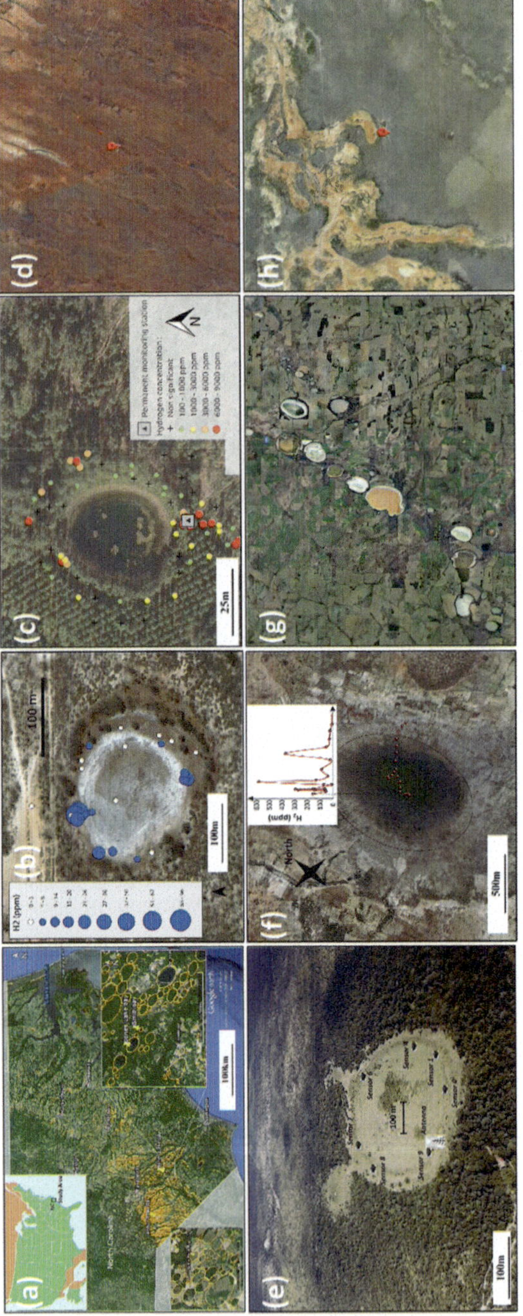

Fig. 2.8 Fairy circles and high hydrogen areas around the world. **a** North Carolina, USA [31]. **b** Perth Basin [102]. **c** The Southern Region of Gironde, France [103]. **d** Meda-1 well, Western Australia [100]. **e** São Paulo Basin, Brazil [91]. **f** Mali, Africa [2]. **g** Yilgarn Craton, Western Australia [100]. **h** Mt Kitty 1 well, Northern Australia [100]

Fig. 2.9 The geological environment of natural hydrogen discovery. **a** Kimberlite [104]. **b** Igneous rocks [105]. **c** Orebodies within a gold deposit [106]. **d** Precambrian rocks [107]. **e** Evaporites [108]. **f** Peridotite [109]

in exploration technology and the need for sustainable energy, has made natural hydrogen a complementary resource in the energy transition. In addition, due to its mantle origin, which is associated with diamonds, can synergize with diamond mining infrastructure and thus increase economic vitality [112]. While research is still in its infancy, preliminary evidence of H_2 seepage and serpentinization potential underscores the need for targeted studies to quantify reserves and optimize mining strategies. As exploration expands, kimberlite-hosted H_2 could become a significant contributor to a low-carbon energy system.

2.5 Igneous Environment

Ophiolites, composed of serpentinized ultramafic rocks, mafic intrusives, basaltic lavas, and marine sediments, originating in active continental marginal areas [40]. Within such igneous settings, H_2 generation is predominantly driven by serpentinization, a water–rock reaction between ultramafic minerals (e.g., olivine, pyroxene) and aqueous fluids. This process occurs under moderate temperatures (200–500 °C) and reducing conditions, yielding serpentine minerals and H_2 [93]. Magmatic degassing and hydrothermal interactions in fractured volcanic–plutonic systems further contribute to H_2 production, particularly in tectonically active zones like mid-ocean ridges, ophiolite complexes, and geothermal fields [119, 120].

H_2 generation efficiency in igneous environments depends on temperature, pressure, mineralogy, and fluid accessibility [120]. Elevated temperatures accelerate serpentinization, Structural permeability, facilitated by faults or fractures, enables fluid infiltration and

H_2 migration. Favorable settings include mid-ocean ridges and ophiolites, where mantle-derived ultramafic rocks interact with seawater [57, 67, 121]; continental rifts and fault zones, which expose mafic–ultramafic rocks to deep groundwater circulation [122]; and volcanic-geothermal systems, where magmatic degassing and hydrothermal activity coexist. Cratonic regions with ancient igneous intrusions may also host H_2 accumulations, particularly where tectonic rejuvenation enhances fluid pathways.

Recent progress highlights the viability of igneous environments as H_2 reservoirs. Pilot projects in Australia, Mali, and Oman target serpentinized ophiolites and rift-related ultramafic terrains, combining drilling campaigns with geochemical assessments to quantify H_2 fluxes. Discoveries of natural seeps in Iceland and mid-ocean ridges underscore the role of tectonically active zones in H_2 migration and entrapment. Concurrently, advances in geochemical modeling and field studies have refined estimates of serpentinization rates and H_2 generation capacities, addressing critical knowledge gaps in resource evaluation. Despite technical challenges in detection and extraction, these developments underscore the growing potential of igneous systems as sustainable H_2 sources [57, 67].

2.6 Volcanic Settings

In volcanic regions, H_2 is released as a constituent of magmatic gases during volcanic eruptions. This H_2 can derive from mantle sources or be generated through reactions occurring at elevated temperatures between interactions of magmatic fluids with aqueous environments. During the ascent of volcanic gases, H_2 may migrate through crustal pathways and accumulate in proximal subsurface geological traps [7].

The key controlling factors include volcanic activity intensity, tectonic proximity, and geothermal conditions. Active volcanic systems exhibit higher H_2 fluxes compared to dormant or extinct counterparts due to sustained magmatic input [120]. Enhanced H_2 emissions are also linked to tectonically active boundaries, where frequent magmatism and fracturing promote gas migration. Additionally, elevated temperatures near volcanic vents optimize H_2 production by accelerating water–rock reactions [97]. These parameters collectively govern spatial and temporal variations in H_2 distribution within volcanic terrains.

2.7 Orebody Environment

The orebody environment refers to the area where minerals are concentrated in Earth's crust through various geological processes. Key geological characteristics include sulfide-rich volcanogenic massive sulfide (VMS) deposits, banded iron formations (BIFs), and uranium ore systems. Ore deposits, or orebodies, often contain metals such as iron, uranium, and sulfide minerals, the high reactivity of these minerals, coupled with structural features such as faults and fractures, facilitates fluid migration and sustained H_2 generation under specific temperature, pressure, and geochemical conditions [97, 123–125].

H_2 production in orebody environments is primarily governed by four mechanisms: (i) water–rock interactions: involving oxidation of reduced iron ($Fe^{2+} + H_2O \rightarrow Fe^{3+} + H_2$); (ii) Sulfide Oxidationor: involving oxidation of sulfides ($MS + H_2O + O_2 \rightarrow M^{2+} + SO_4^{2-} + H_2$); (iii) radiolysis of water induced by radiation from uranium or thorium decay ($H_2O \rightarrow H_2 + O_2$); (iv) deep subsurface reactions in mines or boreholes, where reactive minerals interact with penetrating fluids. Critical factors influencing H_2 yield include the presence of reducing minerals, geological structures enabling fluid migration, and long-term radioactive decay in isolated environments [97].

Current research highlights H_2 seeps near mineralized zones (e.g., Canadian Shield uranium deposits, Scandinavian iron orebodies) and accumulations in deep mines, prompting exploration into commercial extraction. Geochemical studies aim to quantify H_2 fluxes from water–rock reactions and assess accumulation potential. However, challenges persist in distinguishing biogenic from abiogenic H_2, mitigating safety risks in mining operations, and optimizing extraction technologies. Future efforts will focus on integrating geophysical surveys, geochemical modelling, and subsurface monitoring to evaluate the viability of orebody-hosted H_2 as a sustainable energy resource.

2.8 Precambrian Environment

Precambrian environments, comprising ancient cratons, shields, and greenstone belts, host crystalline rocks such as granites, gneisses, and basalts. These stable terrains include deep-seated fractures, iron-rich rocks, and radioactive elements, which are critical for natural hydrogen generation [126]. Low porosity coupled with extensive fracturing enables water–rock interactions at depth, facilitating H_2 production via radiolysis, serpentinization, and redox reactions.

The occurrence of H_2 in these environments is mainly affected by the concentration of radioactive elements, the geological stability of the formation, and the fracture network [6]. H_2 genesis is driven by three principal mechanisms: (i) Radiolysis dominates in uranium-rich granitic terrains, where radioactive decay of uranium, thorium, and potassium dissociates water into H_2 and O_2; (ii) Water–rock interactions in iron-bearing crystalline rocks generate H_2 through oxidation of reduced iron, particularly within deep fracture networks; (iii) Serpentinization, though less common, occurs in Precambrian greenstone belts containing ultramafic remnants, producing H_2 via hydration of olivine-rich lithologies. Notable case studies, such as the Canadian and Fennoscandian Shields, demonstrate H_2 accumulation in uranium-enriched granites and fractured systems, with migration pathways governed by ancient crustal fractures.

Challenges in Precambrian H_2 exploration include deep drilling requirements, low porosity of host rocks, and uncertain trapping mechanisms. However, advances in geophysical imaging and drilling technologies, coupled with stable geological conditions and extensive fracture networks, enhance prospects for commercial viability. As a low-carbon energy resource, natural hydrogen from these ancient terrains could play a pivotal role in global energy transitions, particularly in regions with limited renewables, provided economic and technical hurdles are addressed through targeted research and innovation.

2.9 Salt Deposit Environment

Salt deposits (evaporites), primarily composed of halite, gypsum, and anhydrite, form impermeable seals capable of trapping gases such as H_2 over geological timescales. These deposits are formed through the evaporation of seawater or saline lakes, leading to the precipitation of minerals. These deposits are commonly found in sedimentary basins, salt domes, salt pillows, and salt sheets. Their low permeability and plasticity enable exceptional sealing, preventing gas escape and oxidation. This makes salt deposits analogous to hydrocarbon reservoirs, where H_2 generated in deeper strata migrates upward and accumulates beneath or within evaporite layers. Key factors influencing entrapment include seal integrity, underlying H_2 producing source rocks, and structural features such as folds or faults that enhance storage potential [9].

While salt deposits themselves do not generate H_2, they act as barriers for H_2 produced in adjacent or underlying rocks [9]. Major generation processes include water–rock Interactions in adjacent sedimentary and crystalline rocks (e.g., oxidation of magnetite), radiolysis in radioactive rock layers, and serpentinization in deep ultramafic rocks. H_2 migrates via fractures or permeable pathways into salt-capped reservoirs, where the salt's impermeability and self-sealing plasticity prevent further escape. Critical controls on accumulation include the continuity of salt layers, the productivity of source rocks, and subsurface fracture dynamics, as hydrogen's small molecular size necessitates robust seals to counteract its high mobility.

Case studies in the South Oman Salt Basin, Zechstein Basin, and Gulf Coast salt domes highlight salt deposits as viable H_2 traps, yet exploration faces challenges. Key hurdles involve identifying deep H_2 sources, adapting hydrocarbon-centric detection methods to hydrogen's unique properties, and resolving uncertainties in migration pathways. Economic viability remains unproven, requiring advances in sensor technology and subsurface modeling. However, salt deposits offer strategic advantages, including compatibility with existing hydrocarbon infrastructure and long-term preservation potential. As demand for clean energy grows, these environments may emerge as critical targets for natural hydrogen development, provided technical and economic barriers are addressed through targeted research and pilot projects.

2.10 Ultramafic Rocks

H_2 generation in ultramafic environments predominantly occurs through serpentinization, a process driven by water–rock interactions. When aqueous fluids infiltrate olivine- and pyroxene-rich lithologies such as peridotite, they induce mineral alteration, releasing H_2 as a byproduct. These reactions are particularly prevalent near tectonic plate boundaries, where dynamic geological settings enhance fluid circulation and reactive surface exposure, establishing these zones as critical H_2 sources [97].

The efficiency of serpentinization is governed by multiple parameters. Foremost, rock mineralogy dictates reactivity, with high olivine and pyroxene concentrations accelerating H_2 production [7]. Additionally, sustained water accessibility through interconnected fracture networks is essential to maintain the hydration of fresh mineral surfaces, ensuring prolonged reaction kinetics.

Tectonic activity further modulates H_2 generation by regulating fluid pathways and rock exposure. Active faulting fractures impermeable ultramafic units, enabling deeper fluid penetration and renewing reactive interfaces. This coupling between tectonic stress and fluid influx amplifies serpentinization rates, highlighting the interdependence of structural and geochemical processes in H_2 formation [7].

2.11 Rift Zone

Rift zones have long been recognized as critical environments for the generation and migration of reduced gases, owing to their association with deep-seated tectonic and magmatic processes. While mid-ocean rift systems have been extensively documented as prolific sources of H_2, with numerous studies reporting H_2 emissions from hydrothermal vents and gas-rich fluid inclusions [1]. Onshore rift-related H_2 occurrences remain comparatively rare. This disparity highlights unresolved questions regarding the mechanisms of H_2 migration, preservation, and surface expression in continental rift settings. For instance, the North American Mid-Continent Rift System (MCRS) represents one of the few documented terrestrial examples where soil gas anomalies exceeding 6000 ppm (0.6%) H_2 have been detected in Kansas, spatially correlating with gravity and magnetic anomalies characteristic of the rift's subsurface architecture (Fig. 2.10a). Such findings suggest that continental rifts may host underappreciated H_2 reservoirs linked to deep crustal processes (Fig. 2.10b) [1].

Notably, analytical studies of gas inclusions in minerals from oceanic rift zones reveal remarkably high H_2 concentrations, averaging 21.4%, where serpentinization and water–rock interactions likely drive H_2 generation. The global mid-ocean rift system is estimated to emit approximately 0.12 Tg/year of H_2, a flux considered modest when compared to H_2 production potential in ophiolite-hosted serpentinization systems. This discrepancy raises questions about the completeness of current flux estimations, particularly given the

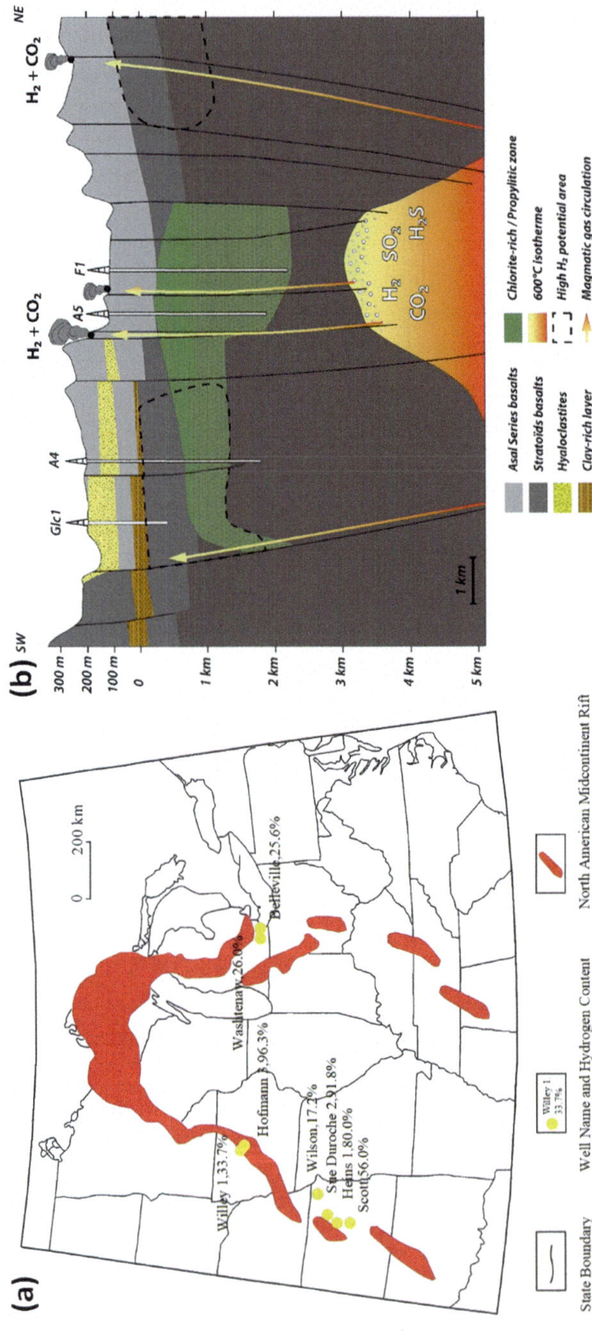

Fig. 2.10 Distribution Models of Natural Hydrogen in Continental Rift Systems. **a** Location of Central Rift Valley and natural hydrogen display in the United States [1, 64]. **b** A conceptual model illustrating the geothermal system and prospective hydrogen accumulation zones within the Asal–Ghoubbet Rift [182]

challenges in detecting and quantifying diffuse H_2 emissions in continental environments [1].

The anomalously high soil gas concentrations observed in the MCRS, coupled with evidence of deep-seated structural controls, imply that continental rifts may contribute non-trivial H_2 fluxes that remain underreported due to limited monitoring and sampling efforts. Future research integrating geophysical, geochemical, and mineralogical approaches will be essential to refine global H_2 budgets and evaluate the resource potential of rift-related systems.

2.12 Sedimentary and Metamorphic Environment

Sedimentary rocks serve as critical environments for natural hydrogen production through multiple geochemical processes. Water–rock interactions, particularly the reduction of ferric iron in iron-rich minerals (e.g., hematite, magnetite), generate H_2 under anoxic conditions. Organic matter decomposition, especially in thermally matured shales, releases H_2 as a byproduct of kerogen breakdown. Radiolysis of water, driven by radiation from uranium, thorium, and potassium in fine-grained sediments, further contributes to H_2 generation over geological timescales [184]. Additionally, serpentinization in underlying ultramafic rocks may produce H_2 that migrates into overlying sedimentary strata via fractures. Key factors influencing these processes include organic content, porosity–permeability architecture, and thermal maturity [7, 97, 120].

Post-generation, H_2 migration in sedimentary basins is governed by rock porosity and permeability. Sandstones facilitate lateral and vertical transport, while structural traps (e.g., anticlines, fault blocks) and stratigraphic traps (e.g., pinch-outs) enable accumulation. Due to hydrogen's minimal molecular diameter and elevated diffusion capacity, the presence of low-permeability barriers, such as shale or evaporite sequences, is essential for effective confinement. Capillary seals in fine-grained rocks are essential to prevent leakage, with evaporites forming particularly robust barriers [185]. Case studies, such as the Bourakebougou field (Mali) and Midwestern US basins, highlight H_2 entrapment in sedimentary reservoirs, often associated with hydrocarbon systems and basement-derived migration pathways [45, 186–188].

H_2 generation in coal basins arises from coupled radiolytic and microbial processes, as exemplified by the Appalachian Basin case study [8]. Concurrently, shallower anaerobic microbial communities facilitate H_2 production through β-oxidation of complex organics, yielding H_2 and methane [189]. Critical controls include: radioactive element concentration, microbial metabolic pathways, and coal permeability. Furthermore, coal's microstructural complexity demonstrates significant gas adsorption capacities [190], suggesting potential for subsurface H_2 storage analogously to CH_4/CO_2 sequestration.

In metamorphic environments, H_2 production is linked to devolatilization reactions under elevated pressure–temperature conditions, releasing water that subsequently participates in H_2 generating redox reactions. Laboratory studies reveal sedimentary rocks, particularly carbonates, exhibit high H_2 adsorption capacities (up to 77.2 cm^3/kg), suggesting potential retention despite their permeability. However, exploration challenges persist: hydrogen's propensity to escape requires advanced detection methods (e.g., magnetotellurics) and rigorous evaluation of seal integrity. Economic viability depends on identifying accumulations comparable to conventional hydrocarbons, necessitating interdisciplinary approaches to address leakage risks and refine subsurface models [8].

References

1. Zgonnik, V. 2020. The occurrence and geoscience of natural hydrogen: A comprehensive review. *Earth-Science Reviews* 203: 103140.
2. Prinzhofer, A., C. S. T. Cissé, and A. B. Diallo. 2018. Discovery of a large accumulation of natural hydrogen in Bourakebougou (Mali). *International Journal of Hydrogen Energy* 43 (42): 19315–19326.
3. Maiga, O., et al. 2023. Characterization of the spontaneously recharging natural hydrogen reservoirs of Bourakebougou in Mali. *Scientific reports* 13 (1): 11876–11876.
4. Meng, Q., et al. 2024. Current status, advances, and prospects of research on natural hydrogen (in Chinese). *Oil & Gas Geology* 45 (5): 1483–1501.
5. Şimşek, E., O. Parlak, and A. H. F. Robertson. 2023. Ion-probe (SIMS) U-Pb geochronology and geochemistry of the Upper Cretaceous Kızıldağ (Hatay) ophiolite: Implications for supra-subduction zone spreading in the Southern Neotethys. *Geosystems and Geoenvironment* 2 (3): 100165.
6. Bourdet, J., et al., Natural hydrogen in low temperature geofluids in a Precambrian granite, South Australia. Implications for hydrogen generation and movement in the upper crust. Chemical Geology, 2023. 638.
7. Giuliani, A., et al. 2023. Genesis and evolution of kimberlites. *Nature Reviews Earth & Environment* 4 (11): 738–753.
8. Boruah, A., P. Alaktaraag, and S. and Singh, Utilization of coal for hydrogen generation. International Journal of Coal Preparation and Utilization, 2024: p. 1–22.
9. Malachowska, A., et al., Hydrogen storage in geological formations—the potential of salt caverns. Energies, 2022. 15(14).
10. Epelle, E. I., et al. 2022. Perspectives and prospects of underground hydrogen storage and natural hydrogen. *Sustainable Energy & Fuels* 6 (14): 3324–3343.
11. Guélard, J., et al. 2017. Natural H_2 in Kansas: Deep or shallow origin? *Geochemistry, Geophysics, Geosystems* 18 (5): 1841–1865.
12. Ward, L. K. 1933. Inflammable gases occluded in the pre-Palaeozoic rocks of South Australia. *Transactions and Proceedings of the Royal Society of South Australia* 57: 42–47.
13. Lollar, B. S., et al. 2014. The contribution of the Precambrian continental lithosphere to global H_2 production. *Nature* 516 (7531): 379–382.
14. Smith, N. J. P., et al. 2005. Hydrogen exploration: A review of global hydrogen accumulations and implications for prospective areas in NW Europe. *Geological Society, London, Petroleum Geology Conference Series* 6 (1): 349–358.

15. Wood, B. L. 1972. Metamorphosed ultramafites and associated formations near Milford Sound, New Zealand. *New Zealand Journal of Geology and Geophysics* 15 (1): 88–128.
16. Li, X., Y. Liu, and J. Wen, Geochemical characteristics of the natural gas from Well Wulong-1, Chuxiong Basin, and its geological significance (in Chinese). Natural Gas Industry, 2002(05): p. 16–19+11.
17. Morrill, P. L., et al. 2013. Geochemistry and geobiology of a present-day serpentinization site in California: The Cedars. *Geochimica et Cosmochimica Acta* 109: 222–240.
18. Vacquand, C., Genèse et mobilité de l'hydrogène dans les roches sédimentaires: source d'énergie naturelle ou vecteur énergétique stockable. 2011.
19. Etiope, G., et al. 2017. Methane and hydrogen in hyperalkaline groundwaters of the serpentinized Dinaride ophiolite belt. *Bosnia and Herzegovina. Applied Geochemistry* 84: 286–296.
20. Coveney, R. M., et al. 1987. Serpentinization and the origin of hydrogen gas in Kansas. *GeoScienceWorld* 71 (1): 39–48.
21. Angino, E., et al. 1984. Hydrogen and nitrogen—Origin, distribution, and abundance, a followup. *Oil & Gas Journal* 82 (49): 142–146.
22. Sano, Y., et al. 1985. Chemical and isotopic compositions of gases in geothermal fluids in Iceland. *Geochemical journal* 19 (3): 135–148.
23. Huntingdon, A., Mount Etna and the 1971 eruption—the collection and analysis of volcanic gases from Mount Etna. Philosophical Transactions of the Royal Society of London. Series A, Mathematical and Physical Sciences, 1973. 274(1238): p. 119–128.
24. Symonds, R.B., et al., Mantle and crustal sources of carbon, nitrogen, and noble gases in Cascade-Range and Aleutian-Arc volcanic gases. 2003, US Geological Survey.
25. Nakamura, H. 1961. Thermal waters and hydrothermal activities in Arima hot spring area, Hyogo Prefecture. *Bulletin of the Geological Survey of Japan* 12 (7): 489–497.
26. Mcelduff, B., Inclusions in chromite from Troodos (Cyprus) and their petrological significance. 1991.
27. Dubessy, J., et al. 1988. Radiolysis evidenced by H_2-O_2 and H_2-bearing fluid inclusions in three uranium deposits. *Geochimica et Cosmochimica Acta* 52 (5): 1155–1167.
28. Milkov, A. V. 2022. Molecular hydrogen in surface and subsurface natural gases: Abundance, origins and ideas for deliberate exploration. *Earth-Science Reviews* 230: 104063.
29. Li, Q., et al. 2024. Occurrence states and transformation of natural hydrogen mechanism (in Chinese). *Geological Survey of China* 11 (05): 9–20.
30. Boreham, C.J., et al., Hydrogen in Australian natural gas: occurrences, sources and resources. the Australian Energy Producers Journal, 2021. 61(1): p. 163–191.
31. Zgonnik, V., et al., Evidence for natural molecular hydrogen seepage associated with Carolina bays (surficial, ovoid depressions on the Atlantic Coastal Plain, Province of the USA). Progress in Earth and Planetary Science, 2015. 2(1).
32. Giardini, A. A., G. V. Subbarayudu, and C. E. Melton. 1976. The emission of occluded gas from rocks as a function of stress: Its possible use as a tool for predicting earthquakes. *Geophysical Research Letters* 3 (6): 355–358.
33. Jiang, F., and G. Li. 1981. Experimental studies of the mechanisms of seismo-geochemical precursors. *Geophysical Research Letters* 8 (5): 473–476.
34. Rezaee, R. 2025. Quantifying natural hydrogen generation rates and volumetric potential in onshore serpentinization. *Geosciences* 15 (3): 112.
35. Ueno, Y., et al. 2006. Evidence from fluid inclusions for microbial methanogenesis in the early Archaean era. *Nature* 440 (7083): 516–519.
36. Kharaka, Y., and J. Hanor. 2003. Deep fluids in the continents: I. *Sedimentary basins. Treatise on Geochemistry* 5: 499–540.

37. Lefeuvre, N., et al. 2022. Natural hydrogen migration along thrust faults in foothill basins: The North Pyrenean Frontal Thrust case study. *Applied Geochemistry* 145: 105396.
38. Han, S., et al. 2024. Hydrogen adsorption characteristics of coal rock and its geological significance (in Chinese). *Journal of China Coal Society* 49 (03): 1501–1517.
39. Mahmoud, M., et al. 2021. A review of geothermal energy-driven hydrogen production systems. *Thermal Science and Engineering Progress* 22: 100854.
40. Wei, Q., et al. 2023. Geological characteristics, formation distribution and resource prospects of natural hydrogen reservoir (in Chinese). *Natural Gas Geoscience* 35 (6): 1113–1122.
41. Zeng, L., et al., Role of geochemical reactions on caprock integrity during underground hydrogen storage. Journal of Energy Storage, 2023. 65.
42. Heinemann, N., et al. 2021. Enabling large-scale hydrogen storage in porous media - the scientific challenges. *Energy & Environmental Science* 14 (2): 853–864.
43. Wang, L., et al., The occurrence pattern of natural hydrogen in the Songliao Basin, P.R. China: Insights on natural hydrogen exploration. International Journal of Hydrogen Energy, 2024. 50: p. 261–275.
44. Han, S., et al. 2021. Genesis and energy significance of hydrogen in natural gas (in Chinese). *Natural Gas Geoscience* 32 (09): 1270–1284.
45. Meng, Q., et al. 2014. Distribution and genesis of hydrogen gas in natural gas. *Petroleum Geology & Experiment* 36 (6): 712.
46. Samantaray, S. S., S. T. Putnam, and N. P. Stadie. 2021. Volumetrics of hydrogen storage by physical adsorption. *Inorganics* 9 (6): 45.
47. Bendall, B. 2022. Current perspectives on natural hydrogen: A synopsis. *MESA Journal* 96: 37–46.
48. Truche, L., T.M. McCollom, and I. Martinez, Hydrogen and abiotic hydrocarbons: molecules that change the world. Elements: An International Magazine of Mineralogy, Geochemistry, and Petrology, 2020. 16(1): p. 13–18.
49. Moretti, I., et al. 2021. Hydrogen emanations in intracratonic areas: New guide lines for early exploration basin screening. *Geosciences* 11 (3): 145.
50. Warr, O., et al. 2019. Mechanisms and rates of ^{4}He, ^{40}Ar, and H_2 production and accumulation in fracture fluids in Precambrian Shield environments. *Chemical Geology* 530: 119322.
51. Truche, L. and E.F. Bazarkina, Natural hydrogen the fuel of the 21st century. E3S Web Conf., 2019. 98: p. 03006.
52. Hao, Y., et al. 2020. Origin and evolution of hydrogen-rich gas discharges from a hot spring in the eastern coastal area of China. *Chemical Geology* 538: 119477.
53. Jin, Z., et al. 2024. Discovery of anomalous hydrogen leakage sites in the Sanshui Basin. *South China. Science bulletin* 69 (9): 1217–1220.
54. Tian, Q., et al. 2024. Natural hydrogen: A new carbon-free energy treasure that cannot be ignored (in Chinese). *Scientific and Cultural Popularization of Natural Resources* 01: 4–11.
55. Charlou, J. L., et al. 2002. Geochemistry of high H_2 and CH_4 vent fluids issuing from ultramafic rocks at the Rainbow hydrothermal field (36°14′N, MAR). *Chemical Geology* 191 (4): 345–359.
56. Kelley, D.S. and F.-G. Gretchen L, Abiogenic methane in deep-seated mid-ocean ridge environments: Insights from stable isotope analyses. Journal of Geophysical Research: Solid Earth, 1999. 104(B5): p. 10439–10460.
57. Proskurowski, G., et al. 2006. Low temperature volatile production at the Lost City Hydrothermal Field, evidence from a hydrogen stable isotope geothermometer. *Chemical Geology* 229 (4): 331–343.
58. Meng, Q., et al. 2021. Geological background and exploration prospects for the occurrence of high-content hydrogen (in Chinese). *Petroleum Geology & Experiment* 43 (02): 208–216.

59. Abrajano, T. A., et al. 1988. Methane-hydrogen gas seeps, Zambales Ophiolite, Philippines: Deep or shallow origin? *Chemical Geology* 71 (1): 211–222.
60. Neal, C., and G. Stanger. 1983. Hydrogen generation from mantle source rocks in Oman. *Earth and Planetary Science Letters* 66: 315–320.
61. Dai, J., et al. 2009. Distribution characteristics of natural gas in Eastern China (in Chinese). *Natural Gas Geoscience* 20 (4): 471–487.
62. Yang, X., D. Liu, and S. Tao, Compositions and implications of inclusions in the typical mantle rocks from East China (in Chinese). Acta Petrolei Sinica, 1999(1): p. 27–31+105+4.
63. Kang, J., et al. 2020. Analysis of geochemical characteristics of hydrogen in fault gases in the north section of Yilan-Yitong fault (in Chinese). *Seismological and Geomagnetic Observation and Research* 41 (04): 111–120.
64. Dou, L., et al. 2024. Global natural hydrogen exploration and development situation and prospects in China (in Chinese). *Lithologic Reservoirs* 36 (02): 1–14.
65. Aiuppa, A., et al., Hydrogen in the gas plume of an open-vent volcano, Mount Etna, Italy. Journal of Geophysical Research: Solid Earth, 2011. 116(B10).
66. Sano, Y., et al. 1993. Origin of hydrogen-nitrogen gas seeps. *Oman. Applied Geochemistry* 8 (1): 1–8.
67. Liu, Q. Y., et al. 2024. Natural hydrogen: A potential carbon-free energy source (in Chinese). *Chinese Science Bulletin-Chinese* 69 (17): 2344–2350.
68. Shangguan, Z., and W. Huo. 2001. The δD value of H_2 escaping from geothermal area in Tengchong Hot Sea and its cause (in Chinese). *Chinese Science Bulletin* 15: 1316–1320.
69. Dai, J., et al. 1994. Geochemical characteristics and isotopic composition of carbon and helium of natural gas in hot springs in some regions of China (in Chinese). *Scientia Sinica (Chimica)* 04: 426–433.
70. Gao, Q. 2004. Volcanic hydrothermal activities and gas-releasing characteristics of the Tianchi Lake region, Changbai Mountains (in Chinese). *Acta Geoscientica Sinica* 03: 345–350.
71. Boreham, C. J., et al. 2021. Hydrogen and hydrocarbons associated with the Neoarchean frog's Leg Gold Camp, Yilgarn craton, western Australia. *Chemical Geology* 575: 120098.
72. Myagkiy, A., I. Moretti, and F. Brunet. 2020. Space and time distribution of subsurface H_2 concentration in so-called "fairy circles": Insight from a conceptual 2-D transport model. *BSGF-Earth Sciences Bulletin* 191 (1): 13.
73. Han, S., et al. 2022. Hydrogen-rich gas discovery in continental scientific drilling project of Songliao Basin, Northeast China: New insights into deep Earth exploration (in Chinese). *Science Bulletin* 67 (10): 1003–1006.
74. Shuai, Y., et al. 2009. Geochemical evidence for strong ongoing methanogenesis in Sanhu region of Qaidam Basin (in Chinese). *Scientia Sinica (Terrae)* 39 (06): 734–740.
75. Woolnough, W. 1934. Natural gas in Australia and new guinea. *AAPG Bulletin* 18 (2): 226–242.
76. Bohdanowicz, C. 1934. Natural gas occurrences in Russia (USSR). *AAPG Bulletin* 18 (6): 746–759.
77. Früh-Green, G. L., et al. 2022. Diversity of magmatism, hydrothermal processes and microbial interactions at mid-ocean ridges. *Nature Reviews Earth & Environment* 3 (12): 852–871.
78. Merdith, A.S., S.E. Atkins, and M.G. Tetley, Tectonic controls on carbon and serpentinite storage in subducted upper oceanic lithosphere for the past 320 Ma. Frontiers in Earth Science, 2019. 7.
79. Jian, H., et al. 2024. Hydrothermal flow and serpentinization in oceanic core complexes controlled by mafic intrusions. *Nature Geoscience* 17 (6): 566–571.
80. Foustoukos, D. I., and W. E. Seyfried. 2004. Hydrocarbons in hydrothermal vent fluids: The role of chromium-bearing catalysts. *Science* 304 (5673): 1002–1005.

81. Menzel, M. D., et al. 2025. Controls of focused fluid release in subduction zones: Insights from experimental dehydration of brucite vein networks in serpentinite. *Contributions to Mineralogy and Petrology* 180 (4): 30.
82. Chogani, A., et al. 2025. Geochemistry of lithospheric aqueous fluids modified by nanoconfinement. *Nature Geoscience* 18 (2): 191–196.
83. Vacquand, C., et al. 2018. Reduced gas seepages in ophiolitic complexes: Evidences for multiple origins of the H_2-CH_4-N_2 gas mixtures. *Geochimica et Cosmochimica Acta* 223: 437–461.
84. Bush, V.A. and V.G. Kaz'min, Crystalline basement and fold complex of the Volga-Ural, Pericaspian, and Fore-Caucasus petroliferous basins. Geotectonics, 2008. 42(5): p. 396–409.
85. Khayuzkin, A., et al. 2025. Geochemistry and formation conditions of the Domanik sediments (Semiluksk Horizon) in the Volga-Ural petroleum province. *Russia. Petroleum Science* 22 (2): 607–626.
86. Karolytė, R., et al. 2022. The role of porosity in H_2/He production ratios in fracture fluids from the Witwatersrand Basin. *South Africa. Chemical Geology* 595: 120788.
87. Boileau, C., et al. 2016. Hydrogen production by the hyperthermophilic bacterium Thermotoga maritima part I: Effects of sulfured nutriments, with thiosulfate as model, on hydrogen production and growth. *Biotechnology for Biofuels* 9 (1): 269.
88. Brovarone, A.V., et al., Subduction hides high-pressure sources of energy that may feed the deep subsurface biosphere. Nature Communications, 2020. 11(1).
89. McGrady, J., et al. 2023. Radiolysis of water at the surface of ZrO_2 nanoparticles. *Radiation Physics and Chemistry* 209: 110970.
90. Albers, E., et al., Serpentinization-driven H_2 production from continental break-up to mid-ocean ridge spreading: Unexpected high rates at the West Iberia Margin. Frontiers in Earth Science, 2021. 9.
91. Prinzhofer, A., et al. 2019. Natural hydrogen continuous emission from sedimentary basins: The example of a Brazilian H_2-emitting structure. *International Journal of Hydrogen Energy* 44 (12): 5676–5685.
92. Morag, N., et al., The origin of plagiogranites: Coupled SIMS O isotope ratios, U–Pb dating and trace element composition of zircon from the Troodos ophiolite, Cyprus. Journal of Petrology, 2020. 61(5): p. egaa057.
93. Ballentine, C. J., et al. 2025. Natural hydrogen resource accumulation in the continental crust. *Nature Reviews Earth & Environment* 6 (5): 342–356.
94. Briere, D., T. Jerzykiewicz, and W. Śliwiński. On generating a geological model for hydrogen gas in the Southern Taoudenni Megabasin. (Bourakebougou Area, Mali). 2016.
95. Ngoma, D.H., Design and development of a community based micro-hydro turbine system with hydrogen energy storage to supply electricity for off-grid rural areas in Tanzania. 2020, Newcastle University.
96. Maiga, O., et al. 2024. Trapping processes of large volumes of natural hydrogen in the subsurface: The emblematic case of the Bourakebougou H_2 field in Mali. *International Journal of Hydrogen Energy* 50: 640–647.
97. Wang, L., et al. 2023. The origin and occurrence of natural hydrogen. *Energies* 16 (5): 2400–2400.
98. Bemis, K., R. Lowell, and A. Farough. 2012. Diffuse flow on and around hydrothermal vents at mid-ocean ridges. *Oceanography* 25: 182–191.
99. Su, Y., et al. 2024. Natural hydrogen exploration: A case study of hydrogen wells in the Mali gas field in Africa and global advances (in Chinese). *Oil & Gas Geology* 45 (5): 1502–1510.
100. Vitaly, V., and R. Reza. 2022. Natural deep-seated hydrogen resources exploration and development: Structural features, governing factors, and controls. *Journal of Energy and Natural Resources* 11 (3): 60–81.

101. Jiang, Z., et al. 2024. Prospects for submarine hydrogen exploration and extraction technologies (in Chinese). *Earth Science Frontiers* 31 (4): 183–190.
102. Emanuelle, F., et al. 2021. Natural hydrogen seeps identified in the North Perth Basin, Western Australia. *International Journal of Hydrogen Energy* 46 (61): 31158–31173.
103. Paul, H., et al., Hydrogen gas in circular depressions in South Gironde, France: Flux, stock, or artefact? Applied Geochemistry, 2021. 127(prepublish): p. 104928-.
104. McNair, B., Kimberlite: a rare, brecciated diamond-bearing mantle rock. Geology Base, 2023.
105. Andrei, M., What are the most common types of rocks? 2024.
106. Bodies, C.O., The challenge to extract gold from complex ore bodies. 2023.
107. Snelling, A., creationist-geology-precambrian. creation.com, 1983.
108. Britannica, evaporite. Britannica.
109. Mindat, Peridotite. Mindat.org, 2016.
110. Russell, J. K., R. S. J. Sparks, and J. L. Kavanagh. 2019. Kimberlite volcanology: Transport, ascent, and eruption. *Elements* 15 (6): 405–410.
111. Lévy, D., et al. 2023. Natural H_2 exploration: Tools and workflows to characterize a play. *Sci. Tech. Energ. Transition* 78: 27.
112. Kjarsgaard, B., et al., A review of the geology of global diamond mines and deposits. 2022. p. 2–102.
113. Фридман, А., Природные газы рудных месторождений. Недра, Москва, 1970.
114. Соколов, В., Геохимия природных газов. Недра, Москва, 1971.
115. Войтов, Г. 1986. Химизм и масштабы современного потока природных газов в различных геоструктурных зонах Земли. *Журнал Всесоюзного химического общества* 31 (5): 533–540.
116. Фомичев, А. 2008. Еще раз о нефтегазопроявлениях в кимберлитовых трубках Якутии. *Геология нефти и газа* 5: 58–64.
117. Дроздов, А., Горно-геологические особенности глубоких горизонтов трубки Удачной. Горный информационно-аналитический бюллетень (научно-технический журнал), 2011(3): p. 153–165.
118. Войтов, Г. and Д. Осика, Водородное дыхание Земли как отражение особеннойгеологического строения и театонического развития ее мегаструктур. Альтернативная энергетика и экология, 1982: p. 7–29.
119. Sleep, N. H., et al. 2004. H_2-rich fluids from serpentinization: Geochemical and biotic implications. *Proceedings of the National Academy of Sciences of the United States of America* 101 (35): 12818–12823.
120. Allen, D. E., and W. E. Seyfried. 2004. Serpentinization and heat generation: Constraints from Lost City and Rainbow hydrothermal systems. *Geochimica Et Cosmochimica Acta* 68 (6): 1347–1354.
121. Holm, N. G., and J. L. Charlou. 2001. Initial indications of abiotic formation of hydrocarbons in the Rainbow ultramafic hydrothermal system, Mid-Atlantic Ridge. *Earth and Planetary Science Letters* 191 (1–2): 1–8.
122. Potter, J. and J. Konnerup-Madsen. A review of the occurrence and origin of abiogenic hydrocarbons in igneous rocks. in Meeting on Hydrocarbons in Crystalline Rocks. 2001. London, England.
123. Nivin, V. A. 2019. Occurrence forms, composition, distribution, origin and potential hazard of natural hydrogen–hydrocarbon gases in ore deposits of the Khibiny and Lovozero massifs: A review. *Minerals* 9 (9): 535.
124. Petersilie, I. Hydrocarbonic gases and bitumens if igneous massifs in the central part of the Kola Peninsula. in Dokl. Akad. Nauk Ukr. SSR. 1958.

125. Nivin, V. 2016. Free hydrogen-hydrocarbon gases from the Lovozero loparite deposit (Kola Peninsula, NW Russia). *Applied Geochemistry* 74: 44–55.
126. Warr, O., M. Song, and B.S. Lollar, The application of Monte Carlo modelling to quantify in situ hydrogen and associated element production in the deep subsurface. Frontiers in Earth Science, 2023. 11.
127. Hosgormez, H., G. Etiope, and M. Yalçin. 2008. New evidence for a mixed inorganic and organic origin of the Olympic Chimaera fire (Turkey): A large onshore seepage of abiogenic gas. *Geofluids* 8 (4): 263–273.
128. Etiope, G., M. Schoell, and H. Hosgörmez. 2011. Abiotic methane flux from the Chimaera seep and Tekirova ophiolites (Turkey): Understanding gas exhalation from low temperature serpentinization and implications for Mars. *Earth and Planetary Science Letters* 310 (1–2): 96–104.
129. Deville, E., and A. Prinzhofer. 2016. The origin of N_2-H_2-CH_4-rich natural gas seepages in ophiolitic context: A major and noble gases study of fluid seepages in New Caledonia. *Chemical Geology* 440: 139–147.
130. Thayer, T. 1966. Serpentinization considered as a constant-volume metasomatic process. *American Mineralogist: Journal of Earth and Planetary Materials* 51 (5–6): 685–710.
131. Yuce, G., et al. 2014. Origin and interactions of fluids circulating over the Amik Basin (Hatay, Turkey) and relationships with the hydrologic, geologic and tectonic settings. *Chemical Geology* 388: 23–39.
132. Zgonnik, V., et al. 2019. Diffused flow of molecular hydrogen through the Western Hajar mountains, Northern Oman. *Arabian Journal of Geosciences* 12 (3): 71.
133. Boulart, C., et al. 2013. Differences in gas venting from ultramafic-hosted warm springs: The example of Oman and Voltri ophiolites. *Ofioliti* 38 (2): 143–156.
134. Войтов, Г. 1990. Восстановленные газы (углеводороды) в породах фундамента Русской плиты. *Бюллетень МОИП* 61: 44–61.
135. Карус, Е., et al., Газы и органическое вещество. Кольская сверхглубокая. 1984.
136. Ikorsky, S., The investigation of gases during the Kola Superdeep borehole drilling (to 11.6 km depth). Geol. Jahrb., D, 1999. 107: p. 145–152.
137. Пиковский, Ю. 1963. Некоторые особенности состава природных газов из нижнекембрийских отложений Иркутского амфитеатра. *Геология и геофизика* 5: 59.
138. Башта, К, В Горбачев, and Л Шахторина. 1991. Задачи и первые результаты бурения Уральской сверхглубокой скважины. *Сов. геология* 8: 51–64.
139. Черепенников, А., Водород в природных газах и, в частности, в эффузивных породах в Ириклинском ущелье на Южном Урале. Сб. посвященный 50алетию научной и педагогической деятельности ВИ Вернадского, АН СССР, М, 1936.
140. Стадник, Е. 1970. О содержании водорода в газах, растворенных в подземных водахсеверо-западного обращения. *Актуальные проблемы нефти и газа* 1: 36–44.
141. Заварицкий, А., Коренные месторождения платины на Урале. Выпуск 108. Геологический комитет, Ленинград. 1928.
142. Лидин, Г., et al. Новые данные о выделениях водородных природных газов из ультраосновных пород. in Докл. АН СССР. 1982.
143. Evans, W.C., N.G. Banks, and L.D. White, Analyses of gas samples from the summit crater. 1980 eruptions Mt. St. Helens, 1980: p. 227–231.
144. Кононов, В. 1983. Геохимия термальных вод областей современного вулканизма. *Труды ГИН АН СССР* 379: 216.
145. Canfield, D. E., M. T. Rosing, and C. Bjerrum. 2006. Early anaerobic metabolisms. *Philosophical Transactions of the Royal Society B: Biological Sciences* 361 (1474): 1819–1836.

146. Соколов, В., Газы земли. Наука, Москва, 1966.
147. Sigvaldason, G. 1966. Chemistry of thermal waters and gases in Iceland. *Bulletin Volcanologique* 29 (1): 589–604.
148. Gilat, A., and A. Vol. 2005. Primordial hydrogen-helium degassing, an overlooked major energy source for internal terrestrial processes. *HAIT Journal of Science and Engineering B* 2 (1–2): 125–167.
149. Куликова, Н., О происхождении природных газов Балейских золоторудных месторождений. Профессорско-преподовательская конференция в МГРИ. 1966.
150. Sato, M., and A. Huntingdon. 1973. The collection and analysis of volcanic gases from Mount Etna: Discussion. *Philosophical Transactions of the Royal Society of London Series A* 274 (1238): 127–128.
151. Флоровская, В., Геохимические основы становления жизни. Вестн. МГУ, сер. геол, 1964(2).
152. Башарина, Л., Эксгаляции побочных кратеров Ключевского вулкана на различных стадиях остывания лавы. 1963.
153. Щека, С., А. Гребенников, and А. Карабцов, Силикатно-металлические хондры как индикаторы флюидного режима игнимбритообразующих расплавов. Исследовано в России: электронный научный журнал, 2009: p. 883–893.
154. Blavoux, B., and J. Dazy. 1990. Caractérisation d'une province à CO_2 dans le bassin du Sud-Est de la France. *Hydrogéologie (Orléans)* 4: 241–252.
155. Shcherbakov, A. V., and N. D. Kozlova. 1986. Occurrence of hydrogen in subsurface fluids andthe relationship of anomalous concentrations to deep faults in the USSR. *Geotectonics* 20: 120–128.
156. Nivin, V. A., et al. 2005. A review of the occurrence, form and origin of C-bearing species in the Khibiny alkaline igneous complex, Kola Peninsula. *NW Russia. Lithos* 85 (1–4): 93–112.
157. Войтов, Г. 1971. О химическом составе газов Кривого Рога. *Геохимия* 11: 1324–1331.
158. Онохин, Ф., Горючие газы Хибинского щелочного массива. Советская Геология, 1959: p. 109–118.
159. Войтов, Г., Газопроявления в рудниках. Рудничная аэрология, Москва: Изд-во АН СССР, 1962: p. 127–44.
160. Куликова, Н.Н., Геохимия газов золоторудных месторождений Забайкалья. 1972: Наука.
161. Кравцов, А., Г. Войтов, and А. Фридман. О содержании водорода в свободных струях в Хибинах. in Докл. АН СССР. 1967.
162. Хитаров, Н, et al. 1979. Газы свободных выделений Хибинского массива. *Советская геология* 2: 62–73.
163. Петерсилье, И.А., Геология и геохимия природных газов и дисперсных битумов некоторых геологических формаций Кольского полуострова. 1964: Наука.
164. Фридман, А., Геологические условия газоносности некоторых медно-колчеданных месторождений.. 1961.
165. Гуревич, М, et al. 1960. Материалы к геохимической характеристике природных газов рудных месторождений. *Труды Института Геологии Рудных Месторождений, Петрографии, Минералогии и Геохимии* 46: 83–91.
166. Гуревич, М., Водород в природных газах и его возможная роль в образовании нефтяных углеводородов. 1967.
167. Перевозчиков, Г. 2011. Водород в недрах Кызылкумов. *Разведка и охрана недр* 2: 35–38.
168. Уханов, А., А. Девирц, and Н. Иванов. Изотопно-легкий водород на Кемпирсае (Южный Урал). in Доклады. АН СССР. 1987.

169. Уханов, А., et al., Водородное дыхание недр: струя изотопно-анамального H2 на Кемпирсайском ультрабазитовом массиве (Южный Урал), in 10й семинар по геохимии магматических пород. 1984. p. 196–197.
170. Sherwood Lollar, B., et al. 1993. Abiogenic methanogenesis in crystalline rocks. *Geochimica et Cosmochimica Acta* 57 (23–24): 5087–5097.
171. Sherwood Lollar, B., et al. 2007. Hydrogeologic controls on episodic H_2 release from Precambrian fractured rocks—energy for deep subsurface life on Earth and Mars. *Astrobiology* 7 (6): 971–986.
172. Newell, K. D., et al. 2007. H_2-rich and hydrocarbon gas recovered in a deep Precambrian well in northeastern Kansas. *Natural resources research* 16 (3): 277–292.
173. Sherwood Lollar, B., et al. 1993. Evidence for bacterially generated hydrocarbon gas in Canadian Shield and Fennoscandian Shield rocks. *Geochimica et Cosmochimica Acta* 57 (23–24): 5073–5085.
174. Goebel, E., et al. 1984. Geology, composition, isotopes of naturally occurring rich gas from wells near Junction City. *Kans. Oil & gas journal* 82 (19): 215–222.
175. Angino, E., et al. Spatial distribution of hydrogen in soil gas in central Kansas, USA. in Geochemistry of gaseous elements and compounds. 1990.
176. Goebel, E. D., et al. 1983. Naturally occurring hydrogen gas from a borehole on the western flank of Nemaha anticline in Kansas. *AAPG Bulletin* 67 (8): 1324–1324.
177. Smith, N., It's time for explorationists to take hydrogen more seriously. First Break, 2002. 20(4).
178. Молчанов, В., Генерация водорода в литогенезе. Новосибирск, Наука, 1981.
179. Savchenko, V. P. 1958. The formation of free hydrogen in the Earth's crust as determinedby the reducing action of the products of radioactive transformations of isotopes. *Geochemistry International* 1: 16–25.
180. Морачевский, Ю, А Самарцева, and А Черепенников. 1937. Газоносность толщи калийных солей Верхнекамского месторождения. *Калий* 7: 24–31.
181. Rogers, G.S., Helium-bearing natural gas. Vol. 121. 1921: US Government Printing Office.
182. Pasquet, G., et al. 2023. Natural hydrogen potential and basaltic alteration in the Asal-Ghoubbet rift, Republic of Djibouti. *BSGF - Earth Sciences Bulletin* 194: 9.
183. Moore, B.J. and S. Sigler, Analyses of natural gases, 1917–85. Vol. 9129. 1987: US Department of the Interior, Bureau of Mines.
184. Liu, Q.Y., et al., Natural hydrogen in the volcanic-bearing sedimentary basin: Origin, conversion, and production rates. Science Advances, 2025. 11(4).
185. Yang, C., and Z. J. Jin. 2025. Hydrogen storage state in clay mineral nanopores: A molecular dynamics simulation study. *International Journal of Hydrogen Energy* 105 (4): 1491–1502.
186. Liu, Q., et al. 2015. Genetic types of natural gas and filling patterns in Daniudi gas field, Ordos Basin. *China. Journal of Asian Earth Sciences* 107: 1–11.
187. Huang, F., Water geochemistry in Songliao Basin (in Chinese). 1999: Beijing:Petroleum Industry Press.
188. Liu, Q., et al. 2017. Effects of deep CO_2 on petroleum and thermal alteration: The case of the Huangqiao oil and gas field. *Chemical Geology* 469: 214–229.
189. Coyle, S., Hydrogen storage potential of the Salina Group, Appalachian and Michigan basins. 2022.
190. Ranathunga, A. S., et al. 2017. A macro-scale view of the influence of effective stress on carbon dioxide flow behaviour in coal: An experimental study. *Geomechanics and Geophysics for Geo-Energy and Geo-Resources* 3 (1): 13–28.

191. Гресов, А., А. Обжиров, and А. Яцук, К вопросу водородоносности угольных бассейнов Дальнего Востока. Вестник Камчатской региональной организации Учебно-научный центр. Серия: Науки о Земле, 2010(1): p. 19–32.
192. Headlee, A. 1962. Hydrogen sulfide, free hydrogen are vital exploration clues. *World Oil* 5: 78–83.
193. Boone, W., Helium-Bearing Natural Gases of the United States: Analyses and Analytical Methods. 1958.
194. United States Geological Survey, The Energy Geochemistry Data Base (EGDB). 2009.
195. Anderson, C.C. and H. Hinson, Helium-bearing natural gases of the United States. Analyses and analytical methods. 1950, Bureau of Mines, Washington, DC (USA); Bureau of Mines, Amarillo, Tex.(USA).
196. Левшунов, П., Водород и его значение в процессах преобразования углеводородов.. 1972.
197. Шорохов, Н., Некоторые новые данные о содержании водорода в осадочных породах, in Геология и Перспективы Нефтегазоносности Некоторых Районов СССР. 1960. p. 264–277.
198. Зингер, А. 1962. Молекулярный водород в составе газа, растворенного в водах газонефтяных месторождений Нижнего Поволжья. *Геохимия* 10: 890–898.
199. Гуревич, М. Обнаружение водорода в газах каменноугольных месторождений Кузбасса. in Доклады Академии наук СССР. 1946. Изд-во Академии наук СССР.
200. Сивак, А. 1962. Газоносность угленосных отложений Норильского района. *Труды НИИГА* 121: 96.
201. Суббота, М, and Н Сардонников. 1968. О генезисе газа, состоящего из азота, окиси углерода и водорода, некоторых межгорных впадин Северного Тянь-Шаня. *Геохимия* 5: 612.
202. Бетелев, Н. 1965. О наличии водорода в составе природного газа на юго-восточном Устюре. *ДАН СССР* 161 (6): 1422–1426.
203. Войтов, Г., et al. Некоторые особенности химического состава глубинных газов по данным Аралсорской сверхглубокой скважины (Прикаспийская впадина). in Доклады Академии наук СССР. 1967. Изд-во Академии наук СССР.
204. Шорохов, Н. and Е. Шишенина, Состав органического вещества и газовой фазы Аралсорской скважины. 1967.
205. Несмелова, З, and Е Рогозина. 1963. К вопросу о водороде в природных газах. *Уголь* 4: 27–35.
206. Осика, Д, Т Янковская, and С Зубик. 2002. Водородная дегазация Земли на современном этапе эволюции и ее связь с сейсмичностью. *Труды Института геологии Дагестанского научного центра РАН* 48: 129–133.
207. Бодунов-Скворцов, Е. 1958. Результаты геохимических исследований в южной части Восточной Сибири. *Геология Нефти* 1: 36–39.
208. Сывороткин, В., Глубинная дегазация Земли и глобальные катастрофы. 2002.
209. Guélard, J. 2016. *Caractérisation des émanations de dihydrogène naturel en contexte intracratonique*. PhD memory: Pierre and Marie Curie University.
210. Deville, E. and A. Prinzhofer, Hydrogène naturel. La prochaine révolution énergétique?: Une énergie inépuisable et non polluante. 2017: Belin Éditeur.
211. Яценко, І, et al. 2012. Леткі компоненти в ендогенних сферулах у зв'язку з проблемою флюїдизатно-експлозивного мантійного рудогенезу. *Мінералогічний збірник* 2: 189–199.

Global H_2 Distribution and Location

3

3.1 Global Distribution: Exploration Locations

The exploration history of natural hydrogen systems reflects the increasing acknowledgment of hydrogen as a viable sustainable energy resource. The field has evolved significantly over time, driven by advances in technology and the urgent need for sustainable energy solutions, transitioning from initial theoretical frameworks to modern pilot initiatives. With expanding research and discoveries, natural hydrogen systems are poised to serve as a critical component in facilitating the global transition to sustainable energy [1].

Mali is the first in West Africa to commercialize natural hydrogen. This breakthrough has sparked a global focus on natural hydrogen system. In recent years, America, Australia and other countries and regions have launched the investigation and exploration of natural hydrogen, showing a vigorous development trend (Fig. 3.1; Table 3.1). China's natural hydrogen research started late, and only three favorable zones for large-scale natural hydrogen accumulation were preliminarily delineated [2].

3.2 China: Three Major Areas

Compared to foreign regions where natural hydrogen has been discovered, China also exhibits geological prerequisites conducive to the generation of high-concentration natural hydrogen accumulations. As a composite continent formed through the amalgamation of multiple tectonic plates, China has experienced multiple episodes of subduction and collisional tectonic movements. The suture zones between these plates host multi-phase ophiolite complexes that serve as significant sources of natural hydrogen generation. Moreover, China hosts an abundance of extensive, deep-seated, and long-lived regional

Y. Yuan et al., *Natural Hydrogen*, Synthesis Lectures on Renewable Energy Technologies, https://doi.org/10.1007/978-3-032-14469-0_3

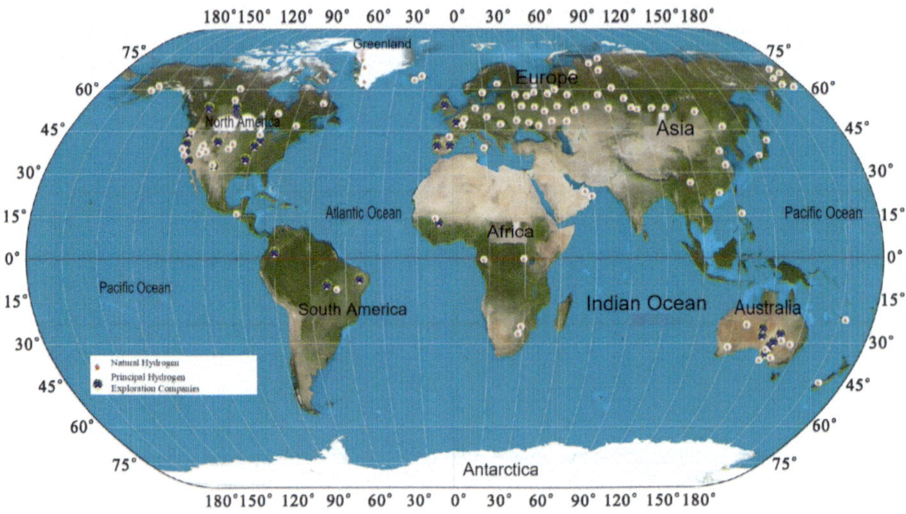

Fig. 3.1 Locations of natural hydrogen resource and exploration [3–8]

fault systems that have remained persistently active [9], including the Tanlu Deep Fault Zone, Altun Deep Fault Zone, and Longmenshan Deep Fault Zone, which provide efficient conduits for the upward migration of deeply derived hydrogen (Fig. 3.2). Additionally, the Cathaysian Rift System and Fen-Wei Rift System share similar geological settings with the North American Rift System, making them promising targets for natural hydrogen exploration [1].

At present, promising natural hydrogen shows have been observed in multiple regions across China (Table 3.2). However, the distribution characteristics of natural hydrogen remain unclear, with limited research on its geological formation conditions, and the distribution and genesis of H_2 in sedimentary basins remain poorly constrained and inadequately explained. Nevertheless, China possesses complete geological prerequisites for natural hydrogen exploration. By superimposing China's major active faults, ophiolite belts, and natural hydrogen occurrences, it has been confirmed that the discovered high-concentration H_2 shows exhibit strong spatial correlation with the favorable geological conditions for H_2 enrichment summarized earlier (Fig. 3.2). H_2 manifestations have been identified in all three types of favorable geological environments. Based on these findings, three prospective zones for natural hydrogen accumulation can be preliminarily delineated: (i) the Tanlu Fault Zone and its surrounding rift basins; (ii) the Altyn Tagh Fault Zone and adjacent basins; (iii) the Sanjiang Tectonic Belt (Nujiang, Lancangjiang, and Jinshajiang) - Longmenshan Fault Zone and associated basin areas [1].

Table 3.1 Some countries explore and develop natural hydrogen projects (modified and translated from [2])

Nation	Relevant parties	Time	Location	Progress
Mali	Hydroma company	2012	Bourakébougou	The hydrogen well in Bourakebougou has commenced operations to provide local electricity
Australia	Gold Hydrogen Company of Australia	2023	Gawler Craton	The first exploratory well in South Australia has discovered natural hydrogen at concentrations of up to 73.3%
United States	CFA Oil Company, USA	1982	Kansas	About 50% hydrogen was found in the Scott well in North America
	NH2E, USA	2019	Kansas	The first natural hydrogen well in the United States was drilled in Kansas
	U.S. Department of Energy ARPA-E	2023	–	$20 million for research and development of clean combustion hydrogen technology in deep rock
	Australian Kelp	2023	Geneva, Nebraska	The company plans to develop the world's first commercial natural ammonia hydrogen well in Geneva, Nebraska, USA
France	Frenchman Ernegrie (FDE)	2023	Lorraine	Natural hydrogen deposits have been discovered in abandoned mines in Lorraine, France, with an estimated natural hydrogen reserve of 46 million tons
	45-8 Energy Company	2023	–	Areas of high potential for natural hydrogen in Europe have been identified

(continued)

Table 3.1 (continued)

Nation	Relevant parties	Time	Location	Progress
Spain	Helios Aragon	2024	Pyrenees	Natural hydrogen is planned to be extracted from old oil wells in Spain in 2024
Brazil	Brazilian Government	2018	São Francisco Basin	A resource assessment was conducted for natural hydrogen escape depressions in the St. Francis Basin and a joint research program was established with the University of Oxford
Korea	Korea National Oil Corporation	2023	–	Announced that natural hydrogen has been discovered at 5 potential sites in South Korea

3.2.1 Tan–Lu Fault Zone and the Peripheral Rift Basin Areas

Under the influence of subduction and collision between the Pacific Plate and Eurasian Plate, eastern China represents a typical back-arc volcanic active zone that has experienced multiple phases of volcanic activity and magmatic intrusions, providing sufficient H_2 sources. As the most significant seismic belt in eastern China, the Tan–Lu Fault Zone has extended into the mantle since the Paleogene period due to compression from the Indian Plate, triggering large-scale basalt eruptions along the fault zone and forming multiple rift basins in surrounding areas. Basins such as Songliao, Bohai Bay, and Subei exhibit well-developed sedimentary reservoir-seal assemblages favorable for the preservation of high-concentration H_2. Currently, numerous high-concentration H_2 shows have been discovered within these basins [1].

3.2.1.1 Songliao Basin: SK-2 Well

Based on an analysis of gas composition data from 1,002 wells in the northern Songliao Basin (see Fig. 3.3a for well distribution), 67 wells (131 samples) exhibit H_2 concentrations exceeding 1%, with 22 wells surpassing 10% (Fig. 3.3b). Hydrogen-bearing gases show H_2 contents ranging from 1.00 to 85.54% (avg. 9.88%, median 4.63%), increasing with depth and peaking in the Yingcheng Formation and basal volcanic reservoirs, whereas CH_4 decreases with depth (Fig. 3.3c). The basement, predominantly granitic, also contains iron-rich intermediate-basic volcanic rocks (e.g., basaltic trachyandesite and basalt, Fig. 3.3d), interpreted as potential deep H_2 sources. H_2 may derive from undiscovered iron-rich mafic rocks or water–rock interactions linked to mantle magmatism.

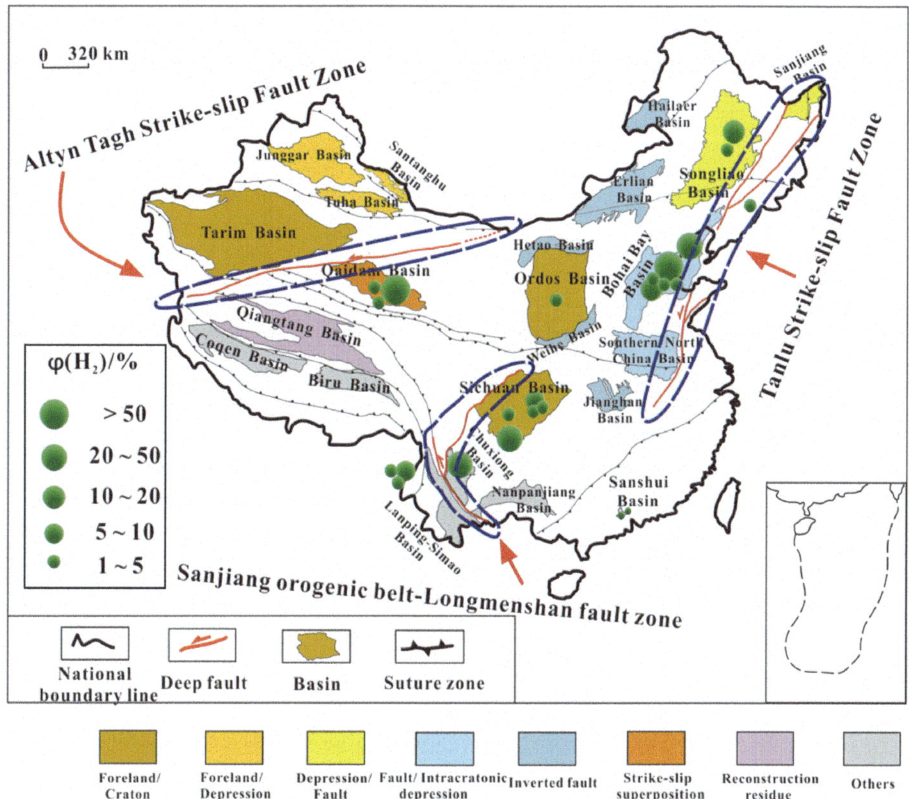

Fig. 3.2 A geographic map summarizing the natural hydrogen resources in China (adapted from [1, 10, 11])

Most high-H_2 formations (> 10% H_2, Table 3.3) are untested via fracturing and occur as dry or low-productivity zones, often underlying conventional alkane gas reservoirs, which dilute H_2 accumulations. Future natural hydrogen exploration should target deep faults near iron-rich basement rocks and avoid areas with active hydrocarbon charging [10].

Table 3.2 Statistics of natural hydrogen display in China (modified and translated from [1])

Basin/area	Gas field/location	φ (H_2)/%	References
Qaidam Basin	Sebei	99.0	[12]
Songliao Basin	SK-2 Well	26.9	[13]
Ordos Basin	Sulige Gas Field	2.1	[14]
East Sichuan	Dukouhe	4.3	[15]
	Luojiazhai	3.5	
		2.6	
	Qianzhong uplift	15.2	[16]
		14.5	
Bohai Bay	Jimo	13.0	[17]
	Dongying Sag	3.9	[18]
		4.5	
		1.0	
		22.8	
		2.8	
		7.5	
Changbai Mountain	Tianchi	1.2	[19]
Tengchong	Rehai	1.1	[20]
		1.2	
		5.2	
Chuxiong Basin	Yanfeng Depression	33.1	[21]
		18.3	
		43.8	
		40.9	
Northern Guizhou	Positive Page 1 Well	24.70–37.0	[22]

The SK-2 Well, located in the northern Xujiaweizi fault depression, intersects multiple formations with significant hydrogen (H_2) occurrences and shows variable concentrations ranging from trace levels (0.0012%) to notably high values (up to 26.89%, Fig. 3.4). H_2 is present across diverse lithologies and does not follow a consistent depth trend; instead, it exhibits a "high–low–high" distribution pattern. Geochemical indicators, including $H_2/^3He$ and R/Ra ratios, reveal a dual crustal and mantle origin for the hydrogen (Fig. 3.5b). Contributions arise from processes such as deep degassing, water–rock interactions, radiolysis, microbial activity, and thermal degradation of organic matter (Fig. 3.5a). In particular, the Shahezi Formation, rich in organic shale and coal, demonstrates that crustal

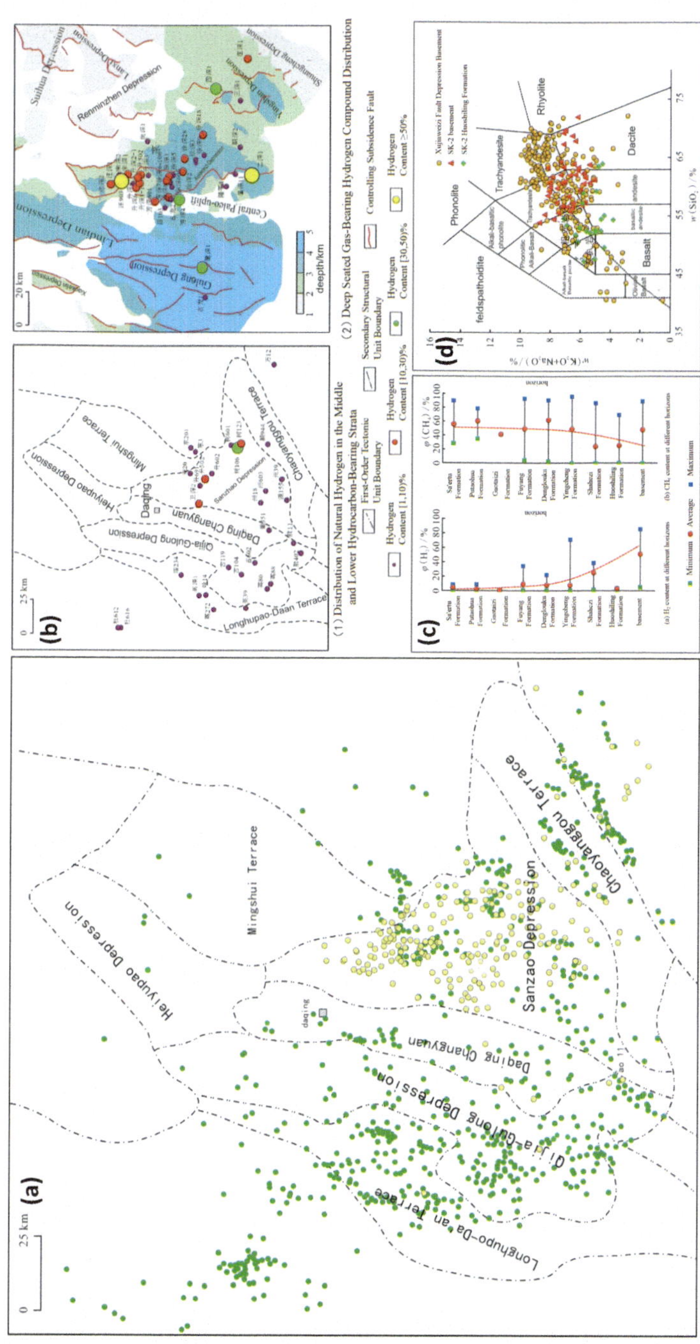

Fig. 3.3 Characteristics and genesis of natural hydrogen in the northern Songliao Basin, Northeast China. **a** Distribution of natural hydrogen survey wells in northern Songliao Basin, Northeast China. **b** Distribution of natural hydrogen content in northern Songliao Basin. **c** Composition characteristics of natural gas in different reservoirs in northern Songliao Basin. **d** TAS classification for basement of Xujiaweizi Rift and volcanic rocks of SK-2 (modified and translated from [10])

Table 3.3 Gas production via well testing and natural gas composition in Songliao Basin, Northeast China (modified and translated from [10])

Number sign	Depth of well testing (m)	Formation (Fm.)	Gas production ($m^3 \cdot d^{-1}$)	φ (H_2) (%)	φ (CH_4) (%)	φ (C_2H_6) (%)	φ (C_3H_8) (%)	φ (CO_2) (%)	φ (CO) (%)	φ (N_2) (%)
Sheng 502	1985.8–2200	Quantou Fm	48,407	0.01	94.41	1.08	0.1	0.18	0	4.15
	2214.6–2223	Quantou Fm		11.10	18.88	0.87	0.08	0.67	13.37	55
Fang shen 801	2966–2986	Denglouku Fm	41,712	0	93.75	1.55	0.55	0	0	4.49
	3075–3081	Denglouku Fm		22.27	2.94	0.39	0.08	5.74	4.73	63.78
Sheng Shen 2–1	2923–2995	Yingcheng Fm	300,789	0.01	92.16	1.5	0.12	2.82	0	3.3
	3132–3152	Yingcheng Fm	881	12.99	33.71	0	0	2.62	17.81	32.87
Wang Shen 1	2989–2998	Yingcheng Fm	202,190	0.01	91.24	1.51	0.22	1.73	0	5.09
	3245–3280	Yingcheng Fm		71.64	9.13	0.14	0	4.31	0	14.78
Fang Shen 4	2823.3–2870	Denglouku Fm	32,656	0	93.44	0.92	0.13	0.42	0	5.04
	3134–3170	Base		39.55	15.03	1.89	2.04	23.86	5.57	11.79
Er Shen 1	3247–3393.5	Base	196	0	87.4	1.34	0.28	0.35	0	10.44
	3375.4–3400	Base		85.54	0	0.74	0	3.24	0	10.48

H_2 can be generated through organic matter pyrolysis, following one of four distinct maturation models (Fig. 3.5c). These findings highlight the complex, multi-source nature of H_2 accumulation in the Songliao Basin [13].

The Yilan–Yitong Fault, trending NNE–NE, constitutes a principal component of the Tanlu Fault Zone north of the Bohai Sea. Along the western branch of the northern segment of the Yilan-Yitong Fault, eight perpendicular profiles spanning from Fangzheng to Luobei were established across confirmed or suspected fault traces (Fig. 3.6a). All survey lines were subjected to systematic observation, with results documented in Fig. 3.6c. Superimposing these survey ranges onto the 2017–2018 seismic epicenter distribution map (Fig. 3.6b) reveals a spatial correlation between elevated H_2 concentrations and zones of heightened modern seismic activity. This geochemical evidence suggests that fault-related

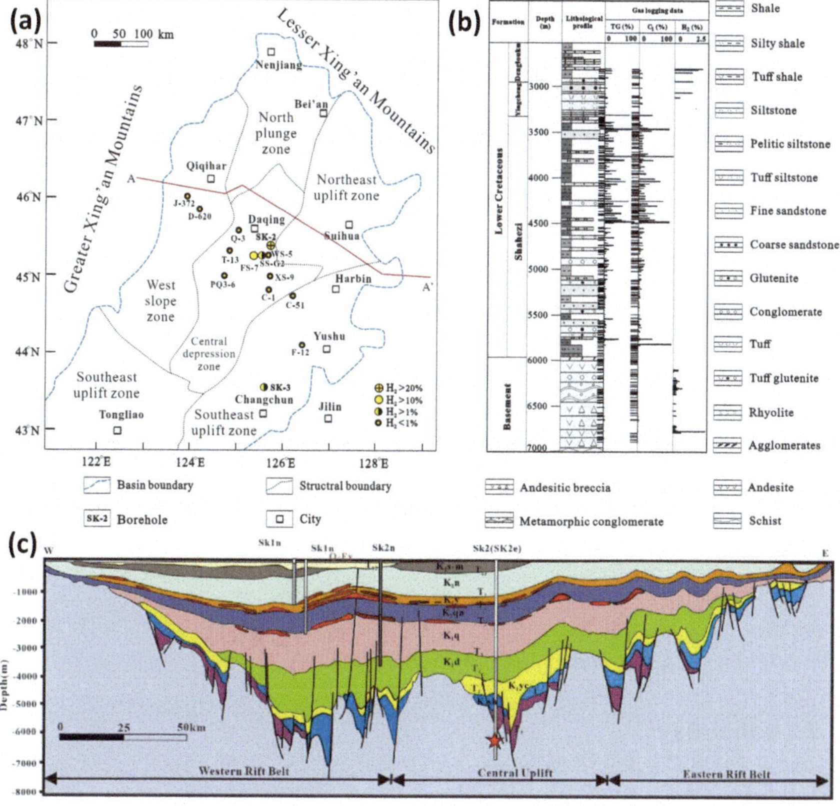

Fig. 3.4 Significant hydrogen-rich gas reservoir identified in the Songliao Basin, northeastern China. **a** The distribution of natural hydrogen gas across various wells. **b** The hydrogen-rich gas logging data of SK-2. **c** stratigraphic profile of Songliao Basin (adapted from [13, 24])

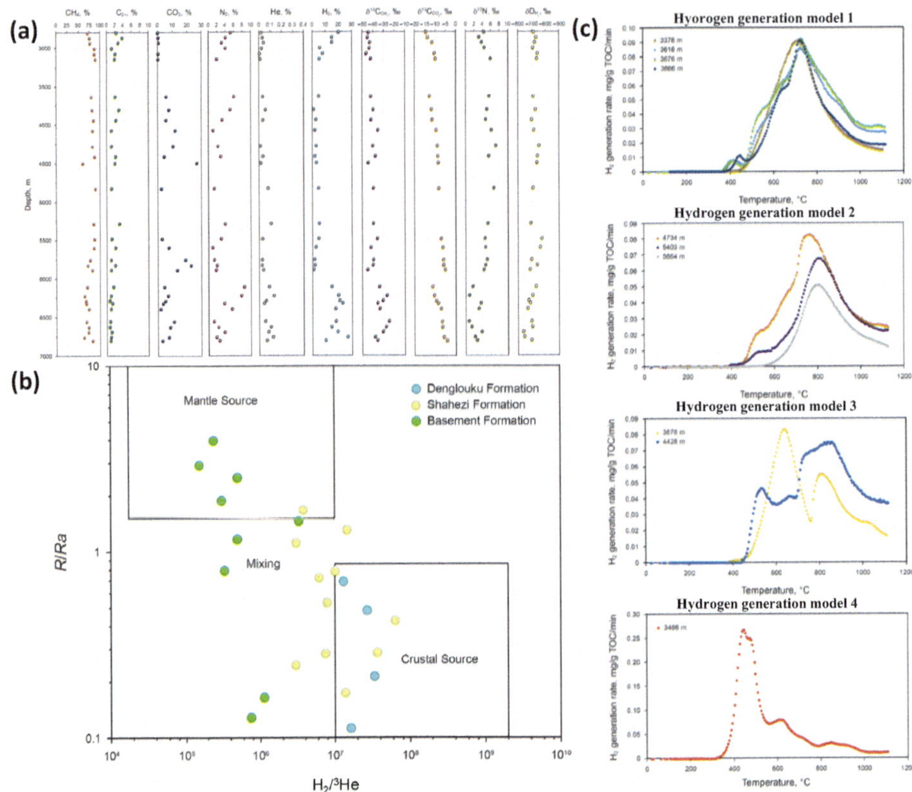

Fig. 3.5 Integrated Geochemical Signatures and Genetic Models of Hydrogen in Natural Gases from the SK-2 Well, Songliao Basin. **a** Variations in natural gas constituents, exemplified by specific isotopic compositions, with respect to depth. **b** Geochemical indicators of hydrogen within natural gas samples from well SK-2. R denotes the $^3He/^4He$ ratio of the sample; Ra represents the atmospheric $^3He/^4He$ ratio. **c** Organic-derived hydrogen genetic mechanisms in the Songliao Basin (reproduced with permission from [25])

H_2 anomalies effectively reflect contemporary tectonic dynamics, potentially serving as an indicator of fault activation status [23].

3.2.1.2 Bohai Bay

In the southeastern Jiaodong Peninsula, a area lies within the NE-trending Muping–Jimo Fault Zone of the Tanlu strike-slip system, characterized by dominant NE and NW fault orientations (Fig. 3.7a–d). Geothermal fluids circulate through fractures associated with these faults, particularly the NW-striking Dongkuang–Shuibo extensional fault, which acts as a major hydraulic conduit. Meteoric recharge from surrounding mountains infiltrates subsurface fractures and reacts with basaltic rocks of the Bamudi Formation (Qingshan

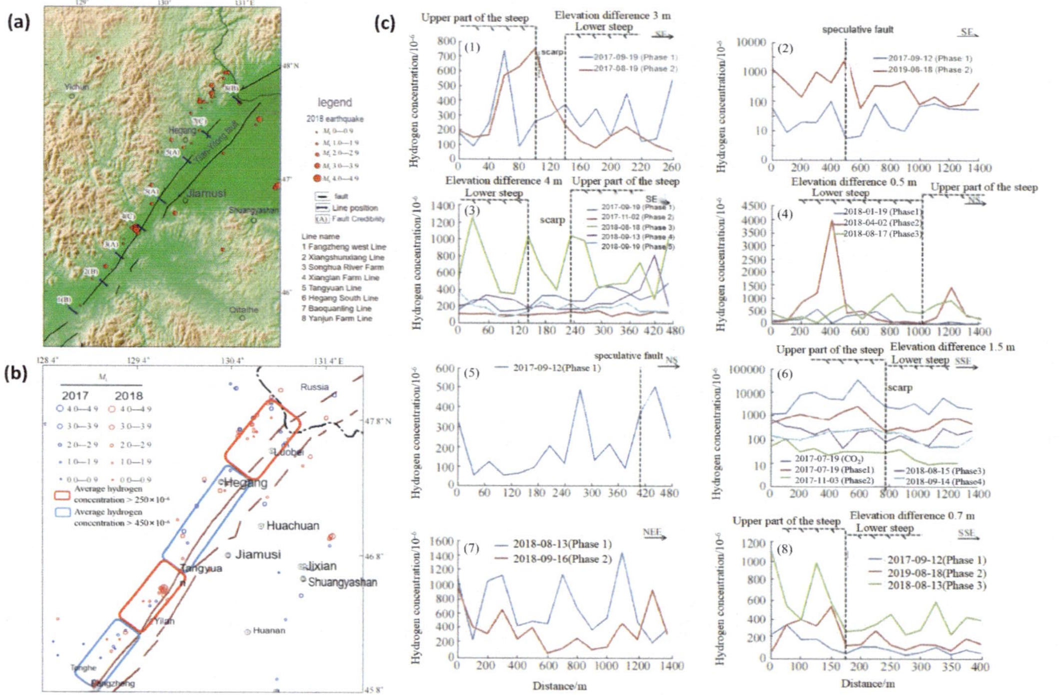

Fig. 3.6 Structural Characteristics, Seismicity, and Hydrogen Anomalies along the Northern Yilan-Yitong Fault Zone. **a** Yilan–Yitong fault structure in the northern section. **b** Seismic distribution of the northern section of Yilan–Itong fault. **c** Correspondence between hydrogen concentration observation curve and fault characteristics. (1) Fangzheng west section; (2) Xiangshunxiang section; (3) Songhua River farm section; (4) Xianglan farm section; (5) Tangyuan section; (6) Hegang South section; (7) Baoquanling section; (8) Yanjun Farm Section (adapted and translated from [23])

Group), approximately 2 km northwest of the geothermal field. This water–rock interaction, involving water reduction and Fe^{2+} oxidation in mafic minerals, generates H_2 at reservoir temperatures of ~ 140 °C, consistent with isotopic evidence (δD: − 836 to − 372‰) (Fig. 3.7e–f). Subsequently, hydrogen is channeled through the NW-oriented fault system and accumulates in sandstone reservoirs, resulting in a conceptual model for gas generation and migration influenced by regional tectonic dynamics and hydrogeological conditions (Fig. 3.7g) [17].

The activity of deep fluids in the Jiyang Depression provides H_2 sources [18]. Deep-derived fluid activities in the Jiyang Depression are distributed along the faults within the basin (Fig. 3.8). Along the Gaoqing–Pingnan Fault Zone, there are deep-derived fluid activities developed over four periods: H_2-rich deep-derived fluids. H_2 is mainly originated from deep-derived fluids near a deep fracture, such as the Gaoqing–Pingnan Fault. Mantle-derived H_2 also exists in areas far away from the deep fracture, that is, in the relatively stable area, such as the Niuzhuang oilfield [26].

The Dongying Sag is a Mesozoic–Cenozoic faulted basin characterized by a northern fault boundary and southern overlap, with steep western slopes and gentle eastern slopes. The Paleogene system constitutes the most significant hydrocarbon source rocks and reservoir strata within the sag. The relatively active tectonic setting during the Cenozoic provided favorable conditions for deep fluid migration. Existing geochemical and geological studies demonstrate that intensive deep fluid activities have occurred in the Dongying Sag [27]. Table 3.4 presents selected geochemical parameters of deep-sourced gases, some of which (determined from fluid inclusions in volcanic rocks) preserve primitive characteristics of deep fluids. As shown in Table 3.4, substantial H_2 content has been detected in deep fluids from the Dongying Sag. For instance, the Well Bin 552 exhibits H_2 concentration exceeding 1%. Considering hydrogen's exceptional mobility, it is inferred that most H_2 has likely been consumed through various reduction reactions [28]. This evidence strongly suggests the historical occurrence of large-scale H_2 activity within the Dongying Sag.

3.2.1.3 Northern Jiangsu and Other Basins

The Huangqiao region is situated in Jiangsu Province, East China, located geographically along the southern flank of the Subei Basin. Structurally, it forms part of the Huangqiao composite synclinal structure in the northeastern Nanjing Depression (Fig. 3.9a). From the Late Yanshan to Early Himalayan tectonic stages, the Subei Basin underwent substantial tectonic subduction and uplift, accompanied by multiple episodes of basic magmatic emplacement, including fissure-type eruptions of tholeiitic and alkaline olivine basaltic lavas. The Tanlu Fault, a significant structure of crustal magnitude that plays a key role in the evolution of the basin, traverses more than 500 km within a fault zone 20–30 km wide and displays sub-vertical dip characteristics. This fault penetrates Neogene strata and extends to the lithospheric mantle in the Subei Basin, reflecting its deep-seated lithospheric structure [30].

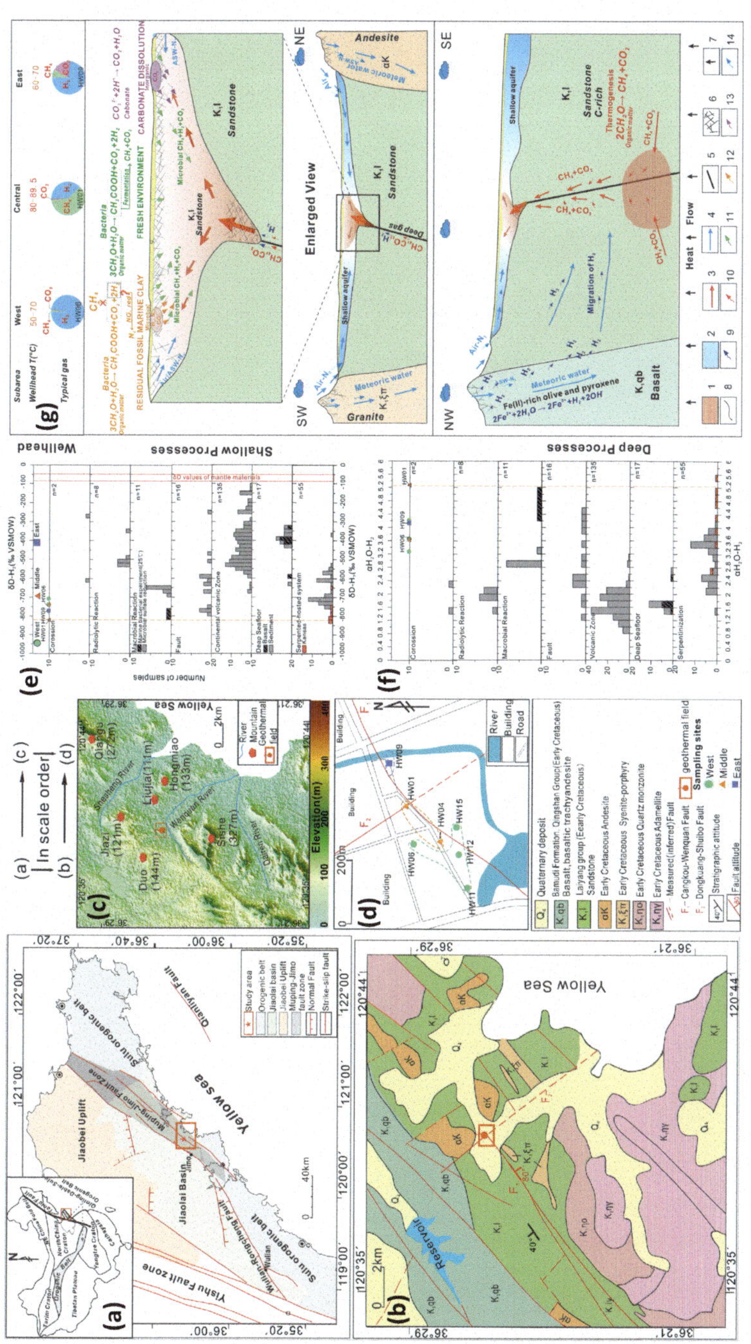

Fig. 3.7 Summary of geological context, sampling, hydrogen isotopic data, and conceptual genetic model for the study area (the southeastern Jiaodong Peninsula). **a** Tectonic setting of the study area. **b** Generalized geological map of the study area. **c** Topographic map of the study area. **d** Sampling locations for geothermal water and gas. **e** Compilation of δD-H_2 values from various systems. **f** Compilation of αH_2O–H_2 data from multiple systems. **g** Conceptual model illustrating mechanisms of gas generation and evolution (reproduced with permission from [17])

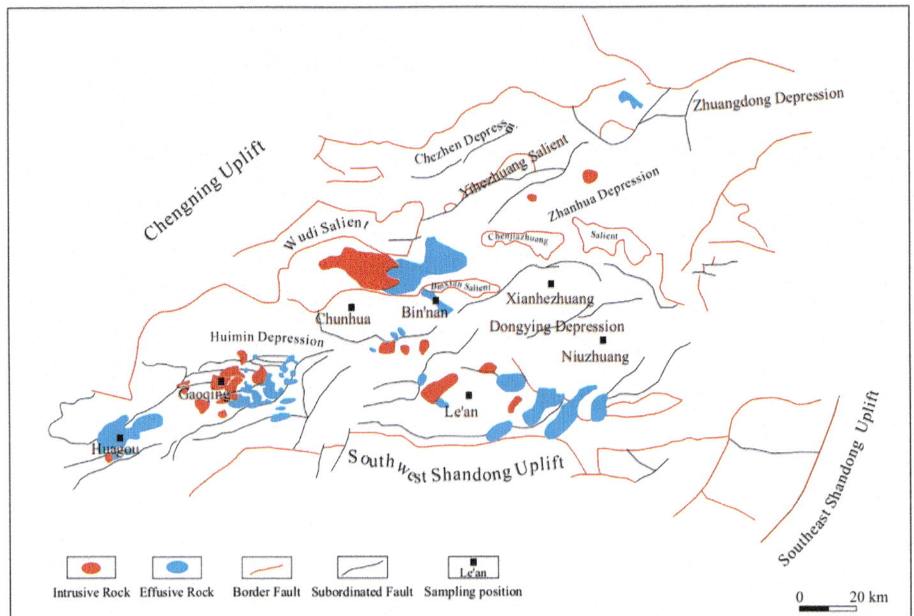

Fig. 3.8 Regional structure of Gaoqing-Pingnan fault zone, Jiyang depression, Shandong Province (adapted and translated from [29])

Table 3.4 Geochemical characteristics of mantle source gas in Dongying Sag (adapted and translated from [27])

Well	Fm.	CO_2 (%)	H_2 (%)	$^3He/^4He$ (10^{-6})	$\partial^{13}C_{CO_2}$ (PDB, ‰)
Bin 4–6-6	Es_4	72.5	3.87 ± 0.11	− 4.57	–
Ping 14–3	Es_4	77.93	4.47 ± 0.14	− 4.32	–
Bin 552	Ed	1.27	1.02	–	–
Fan 115	Es_3	3.19	0.41	–	–
Hua 17	Es_3	93.54	–	4.49 ± 0.12	− 3.35
Gaoqi 3	Ng	94.36	–	4.49 ± 0.12	− 4.41
Gaoqi 2[a]	K_1	–	22.8	–	–
Gao 41–1[a]	K_1	54.3	2.8	–	− 14.21
Yanggu 4[a]	K_1	51.4	7.5	–	–

[a] Indicates volcanic fluid inclusions

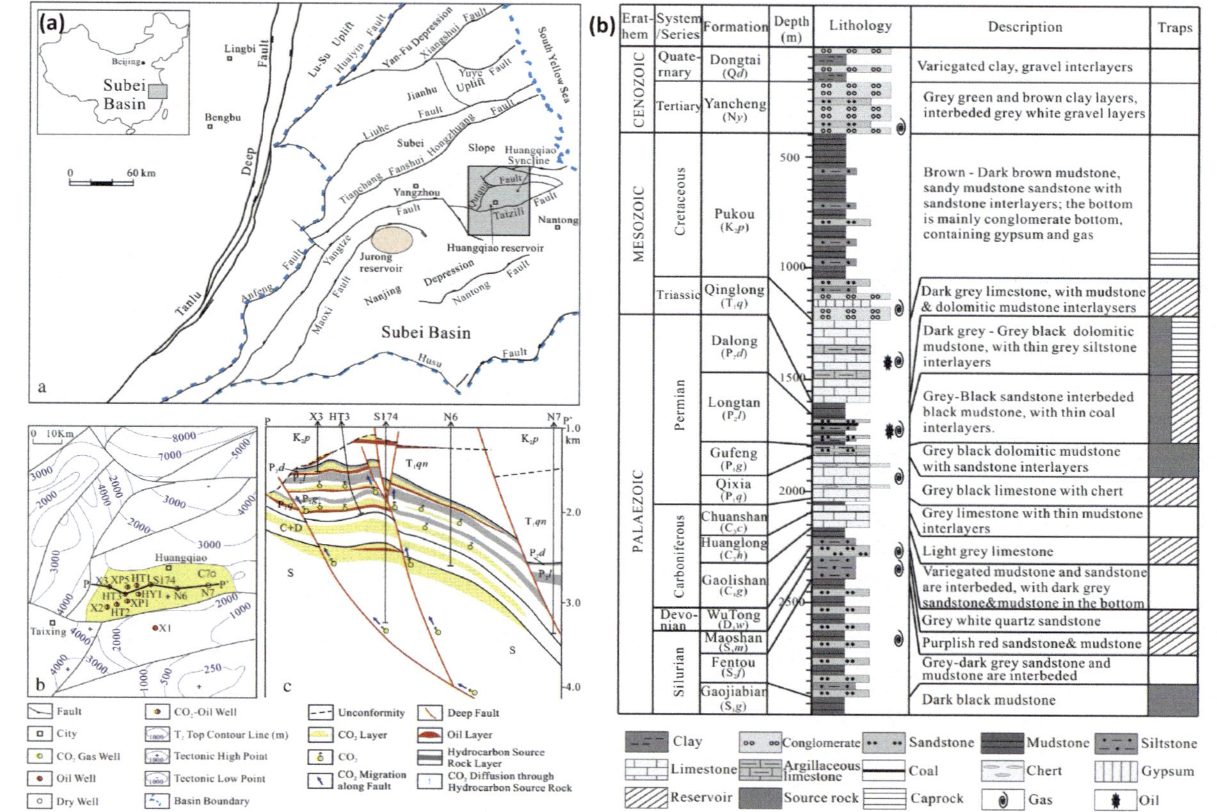

Fig. 3.9 Tectonic diagram and stratigraphic columns of the Huangqiao oil and gas field. **a** Tectonic setting of the Huangqiao gas field in northern Jiangsu. a: Structural framework of the Subei Basin; b: spatial distribution of oil and gas accumulations in the Huangqiao field; c: cross-sectional view of the gas reservoir. **b** Stratigraphic profile of the Huangqiao oil and gas field (reproduced with permission from [30])

The Huangqiao region is predominantly overlain by sedimentary deposits. The oldest stratigraphic unit encountered in exploration is the Lower Silurian Gaojiabian Formation (S1g), with drilling operations yet to penetrate its basal contact (Fig. 3.9b). A notable stratigraphic discontinuity separates the Upper Devonian Wutong Formation (D3w) from the underlying Upper Silurian Maoshan Formation (S3m). Geochemical analyses reveal that natural gas extracted from the Huangqiao gas field in the Subei Basin exhibits H_2 concentrations ranging from 0 to 4.26%, with a mean value of 0.53% [30].

3.2.2 Altyn Fault Zone and the Basin Areas

Western China has undergone multiple subduction-collision tectonic events. Along both sides of the Altun Fault Zone, there developed ophiolite belts of different periods, where identifying concealed ophiolite belts holds key significance for natural hydrogen exploration in this region. The northern Tarim Basin, adjacent to the Altun Fault Zone witnessed extensive magmatic activity during the early Permian, forming the renowned Tarim Large Igneous Province (TLIP). The basin contains mafic–ultramafic dikes that serve as high-quality H_2 sources. In the south of the fault zone, the Qaidam Basin contains substantial potash salt and brine deposits, which favor the generation of hydrogen-enriched fluids [1].

3.2.2.1 Tarim Basin

The Tarim Large Igneous Province (TLIP) covers an area of 200,000–300,000 km^2 (as shown in Fig. 3.10) and contains extensive Permian volcanic and intrusive rocks. The volcanic sequence is predominantly composed of basalts, with minor occurrences of rhyolite, andesite, tuff, and other volcanic lithologies [31]. The TLIP exhibits two distinct magmatic events at ~ 290 and ~ 280 Ma. The earlier phase at approximately 290 Ma is marked by bimodal volcanic assemblages comprising basalt and rhyolite, which are mainly located in the interior region of the Tarim Craton. The subsequent ~ 280 Ma phase comprises intrusive complexes and mafic dike swarms, mainly occurring along the margins of the Tarim Basin. Notably, the basin hosts mafic–ultramafic dikes that serve as high-quality H_2 sources [32].

3.2.2.2 Qaidam Basin: Sebei

The Sanhu Depression in the Qaidam Basin, represents a globally renowned Quaternary biogenic gas province. Gas samples were obtained from production well XS3-4 (a developmental well) in the Sebei-1 Gas Field and reference well SN2 located in the Senan structural belt, in the Sanhu region of the eastern Qaidam Basin (as shown in Fig. 3.11a). Analytical data reveal exceptionally high H_2 concentrations in this region. The absence of deep-seated faults in the Sanhu area precludes upward migration of mantle-derived gases through deep fracture systems. Notably, samples exhibiting elevated H_2 concentrations in headspace gases correspond to high-response zones during gas logging (Fig. 3.11b).

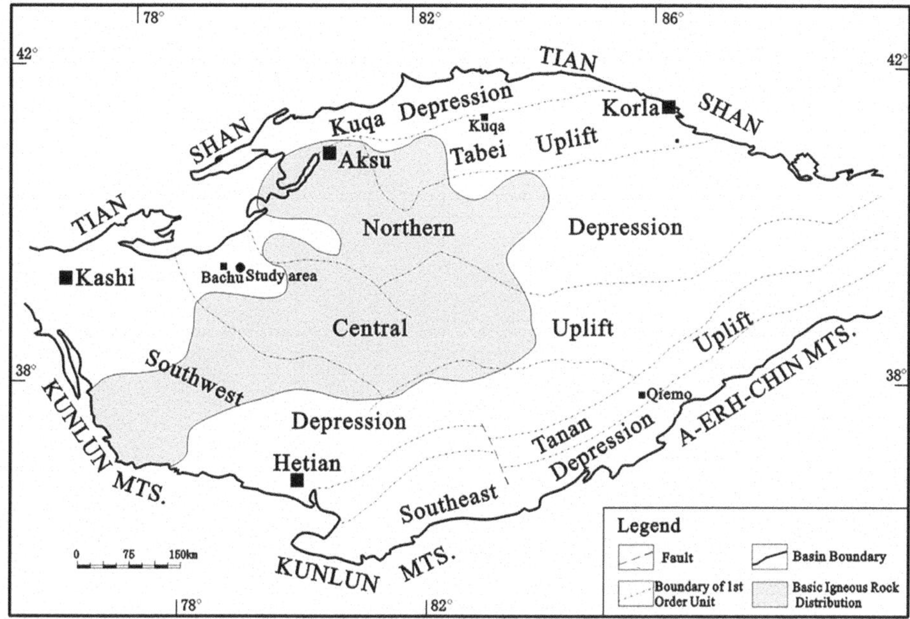

Fig. 3.10 Geological map of Tarim Basin (modified and translated from [33])

These observations collectively suggest that the H_2 in this region originates primarily from sediment desorption rather than subsequent contamination processes [12].

3.2.3 Sanjiang Orogenic Belt–Longmenshan Fault Zone and Peripheral Basin

The Sanjiang region, located at the convergent margin of the Indian and Eurasian plates, is characterized by numerous thrust–strike-slip fault systems and intervening crustal blocks (Fig. 3.12). Since the Himalayan period, intense magmatic–tectonic activity has produced numerous ophiolitic mélange belts and volcanic sequences, offering substantial H_2 sources for natural hydrogen generation. Major active faults serve as effective migration pathways for deep-sourced H_2, with natural hydrogen concentrations exceeding 5% detected in spring gases from Tengchong, Yunnan. Extensional basins in western Yunnan (e.g., Tengchong–Lianghe–Longchuanjiang), formed through Miocene block rotation and pull-apart processes, contain Pliocene–Quaternary fluvial–lacustrine deposits interbedded with basalts. This setting shares similarities with the Mali Bourakebougou natural hydrogen system, indicating high potential for H_2 accumulation. The Longmenshan Fault Zone, thrusting eastward over the Sichuan Basin, provides further favorable conduits for H_2 migration. Combined with the Permian Emeishan Large Igneous Province as a significant

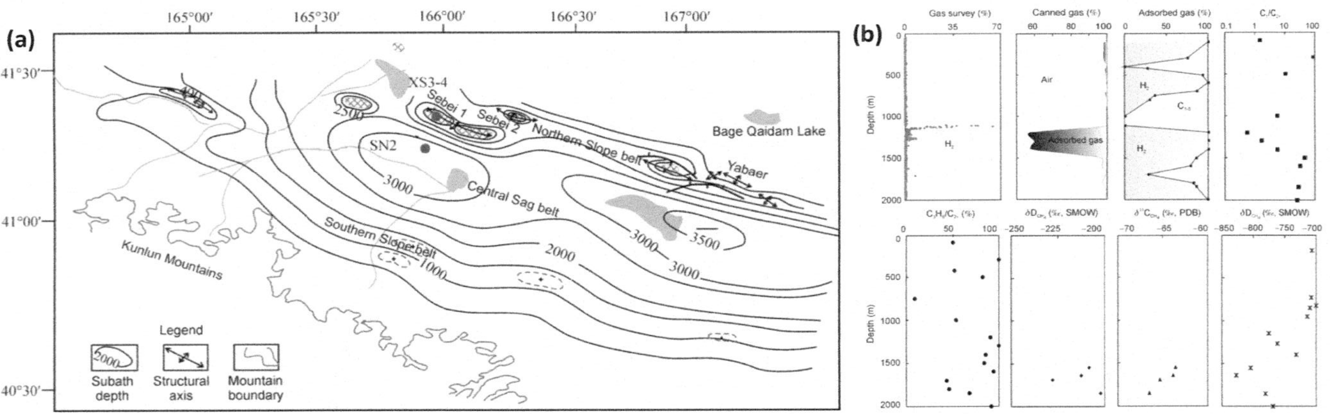

Fig. 3.11 Tectonic-Sedimentary Background and Natural Gas Geochemical Characteristics of the Sanhu Region, Qaidam Basin. **a** the tectono-sedimentary setting in the Sanhu area, Qaidam Basin. **b** Compositional characteristics of headspace gases from well SN2 (adapted and translated from [12])

H_2 source, several eastern basins (e.g., Sichuan, Qianzhong, Chuxiong) show promising natural hydrogen occurrences [1].

3.2.3.1 Tengchong, Yunnan Province

The Tengchong microplate, where the Tengchong volcanic zone is situated, lies between the Indian Plate and Eurasian Plate. It is bounded to the west by the Myitkyina–Mandalay Suture Zone separating it from the Myanmar Block, and to the east by the Nujiang Suture Zone demarcating its boundary with the Baoshan Terrane. The distribution of volcanic rocks and craters in this region shows close genetic relationships with neotectonic activities. Notably, volcanic craters with recent eruptive histories are predominantly aligned along the NS-trending Ruichuan–Tengchong Fault (Fig. 3.13). Current geochemical investigations have detected natural hydrogen with volume fractions exceeding 5% in volcanic geothermal gases from the Tengchong area [20].

3.2.3.2 Chuxiong Basin, Yunnan Province

The Wulong 1 Well is situated at the high structural position of the Wulongkou Anticline within the Yanfeng Sag, northern Chuxiong Basin, demonstrating favorable hydrocarbon source rocks. The thickly-bedded, dark-colored mudstone successions are predominantly deposited in semi-deep to deep lacustrine facies, and are characterized primarily by mixed-type organic matter. Compositional analyses were conducted on natural gas samples collected from two intervals: the open-hole section at 3140–4620 m (Upper Triassic) and the 2411.3–2425 m interval (Fengjiahe Formation). The analytical results are presented in Table 3.5. Notably, natural hydrogen with volume concentrations exceeding 40% has been identified in the Chuxiong Basin [21].

3.2.3.3 Sichuan, Qianzhong

The Sichuan Basin, located in the western Yangtze Paraplatform, represents a NE-trending rhombic-shaped structural and sedimentary petroliferous basin (Fig. 3.14c). It constitutes a large composite petroliferous basin characterized by dual depositional systems: marine carbonate sequences spanning from the Edicarian to Middle Triassic, and terrigenous clastic successions extending from the Late Triassic to Eocene. The Feixianguan Formation gas reservoirs in northeastern Sichuan represent one of the largest gas field clusters in China with relatively high H_2 sulfide concentrations discovered to date. Non-hydrocarbon gases such as nitrogen, helium, and H_2 are also present in these reservoirs. Notably, the Luojiazhai gas field exhibits a maximum H_2 concentration of 3.45%, while the Dukouhe gas field demonstrates an elevated H_2 content reaching 4.31% in compositional analysis [15].

The southeastern Sichuan region is situated within the southeastern depression-fold belt of the Sichuan Basin, adjacent to the basin margin, spanning both the steep structural belt of eastern Sichuan and the moderate-low gentle fold belt of southern Sichuan. Its regional tectonic setting occupies the southeastern lower slope of the Caledonian Leshan–Longnüsi

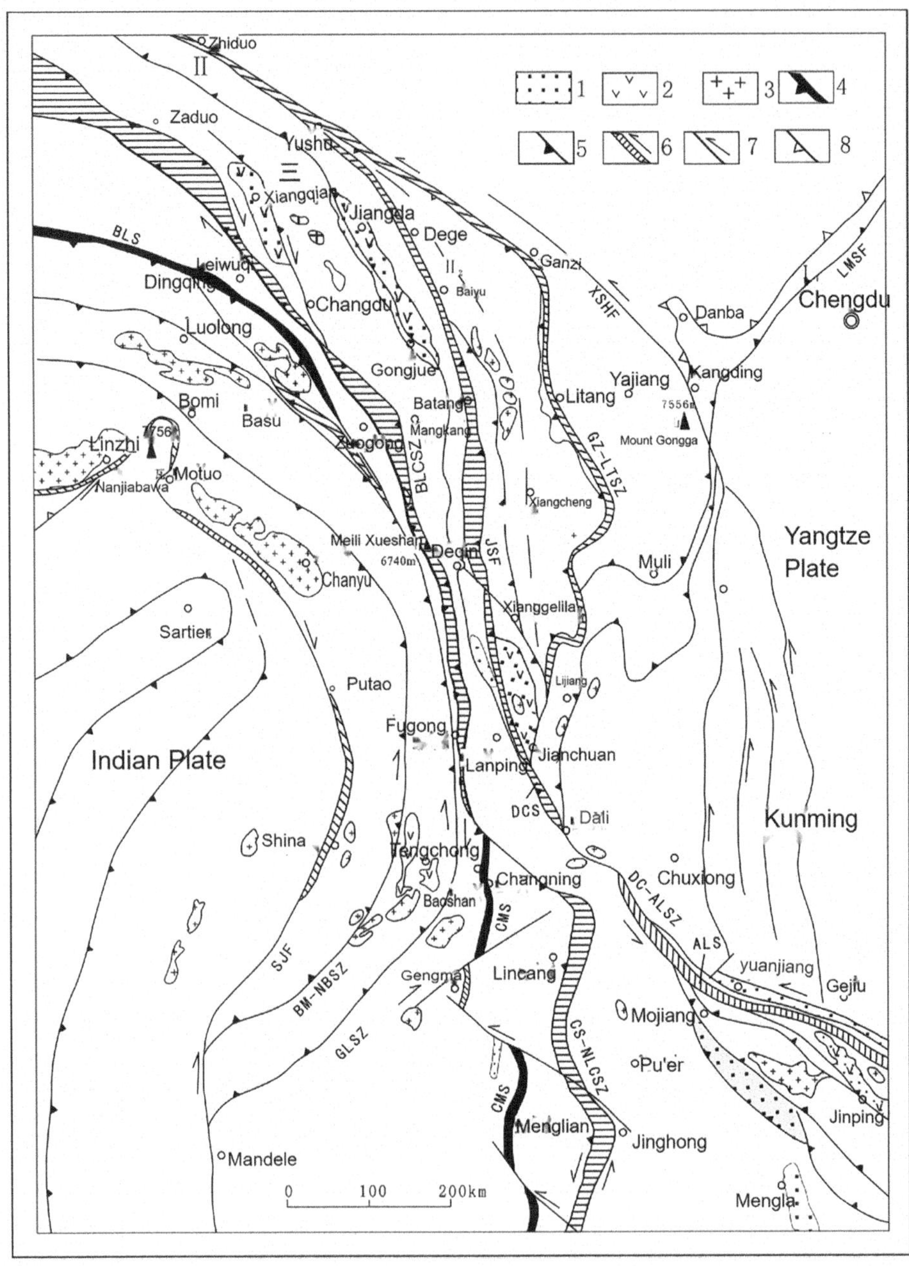

Zhiduo
Zaduo
Yushu
Xiangqian
Jiangda
Dege
Ganzi
BLS
Leiwuqi
Dingqing
Changdu
Baiyu
XSHF
LMSF
Chengdu
Danba
Luolong
Gongjue
Kangding
Yajiang
Litang
Batang
Mangkang
7556m
Mount Gongga
Bomi
Basu
Zuogong
Linzhi
7756
Motuo
Nanjiabawa
BLCSZ
GZ-LTSZ
Xiangcheng
Meili Xueshan
6740m
Deqin
JSF
Muli
Chanyu
Yangtze
Plate
Xianggelila
Sartier
Lijiang
Putao
Fugong
Jianchuan
Lanping
Indian Plate
DCS
Dali
Kunming
Shina
Tengchong
Changning
Chuxiong
Baoshan
DC-ALSZ
CMS
ALS
SJF
yuanjiang
BM-NBSZ
Gengma
Lincang
Gejiu
Mojiang
CS-NLCSZ
GLSZ
Pu'er
Jinping
CMS
Menglian
Jinghong
Mandele
Mengla
0 100 200km

◂**Fig. 3.12** Schematic map of neotectonic structure and Cenozoic basin, magmatic rock and nuclear complex distribution map of Sanjiang Orogenic Belt in Southwest China [34]. *Notes* 1—New generation basins; 2—New generation volcanic rocks; 3—New generation intrusive rocks; 4—Panel suturing tape; 5—Plate subduction zone; 6—Cut away the slip band; 7—Subduction fault; 8—Break off the fault. SJF—Shijie fracture; BM-NBSZ—Bomi–Nabong shear zone; GLSZ—Goligong Mountain slip shear zone; BLS—Bhagong Lake–Nujiang plate suturing zone; CMS—Changning–Menglian plate suturing zone; BLCSZ—North Lancang River strike–slip shear zone; CS-NLCSZ—Chongshan–Lancang River strike–slip shear zone; JSSZ—Jinsha River strike-slip shear zone; DC-AlSZ—Diancangshan–Ailao Mountain slip shear zone; GZ-LTSZ—Ganzi–Litang walking slip shear zone; XSHF—Xianshuihe Fault Zone; LMSF—Longmenshan Fault Zone

paleo-uplift, the southeastern slope of the Indosinian Luzhou paleo-uplift, and the northern marginal lower slope of the Duyunian period Qianzhong paleo-uplift (Fig. 3.14a). H_2 with maximum component content reaching 15.24% has been detected in wells distributed across major structural units of this area [16].

In the northern Guizhou region, folds and faults are well-developed (Fig. 3.14b), with relatively complete stratigraphic sequences from the Edicarian to Silurian systems. Based on lithologic attributes, sedimentary facies distribution, and fossil assemblages, the lower Cambrian Niutitang Formation in this area can be divided into eastern and western facies regions: the western facies region primarily consists of shallow-water shelf deposits, while the eastern facies region is dominated by deep-water shelf deposits. Gas-bearing property analysis of organic-rich mudstones in the Niutitang Formation reveals favorable gas content characteristics. Taking Well Zhengye 1 as an example, which has undergone comprehensive gas chromatography analysis, the hydrogen content (H_2) ranges from 24.70 to 36.98% [22].

3.2.3.4 Sanshui Basin

The Sanshui Basin is a rift basin that developed along the continental margin of South China from the Late Cretaceous through the Miocene. Underlying this basin, the lithosphere exhibits the thinnest profile recorded across mainland China (Fig. 3.15a). Comprehensive field investigations incorporating soil-gas analyses were systematically conducted using portable gas analyzers targeting H_2, CO_2, and Rn concentrations. Analytical data reveal anomalously elevated H_2 concentrations across multiple fault systems, with peak detected levels reaching 6948 ppm (Fig. 3.15b) [11].

Considering the geological framework and soil gas geochemical signatures within the Sanshui Basin, three plausible mechanisms have been hypothesized for the observed H_2. The primary origin may stem from mantle-sourced magmatic degassing. A secondary mechanism could involve crustal-scale water–rock interactions, notably serpentinization of mafic and ultramafic lithologies (Fig. 3.15c). Additionally, $\partial^{13}C_{CO_2}$ values in soil gases, some falling below − 10‰, support a tertiary origin attributable to organically derived H_2 via thermal degradation processes [11].

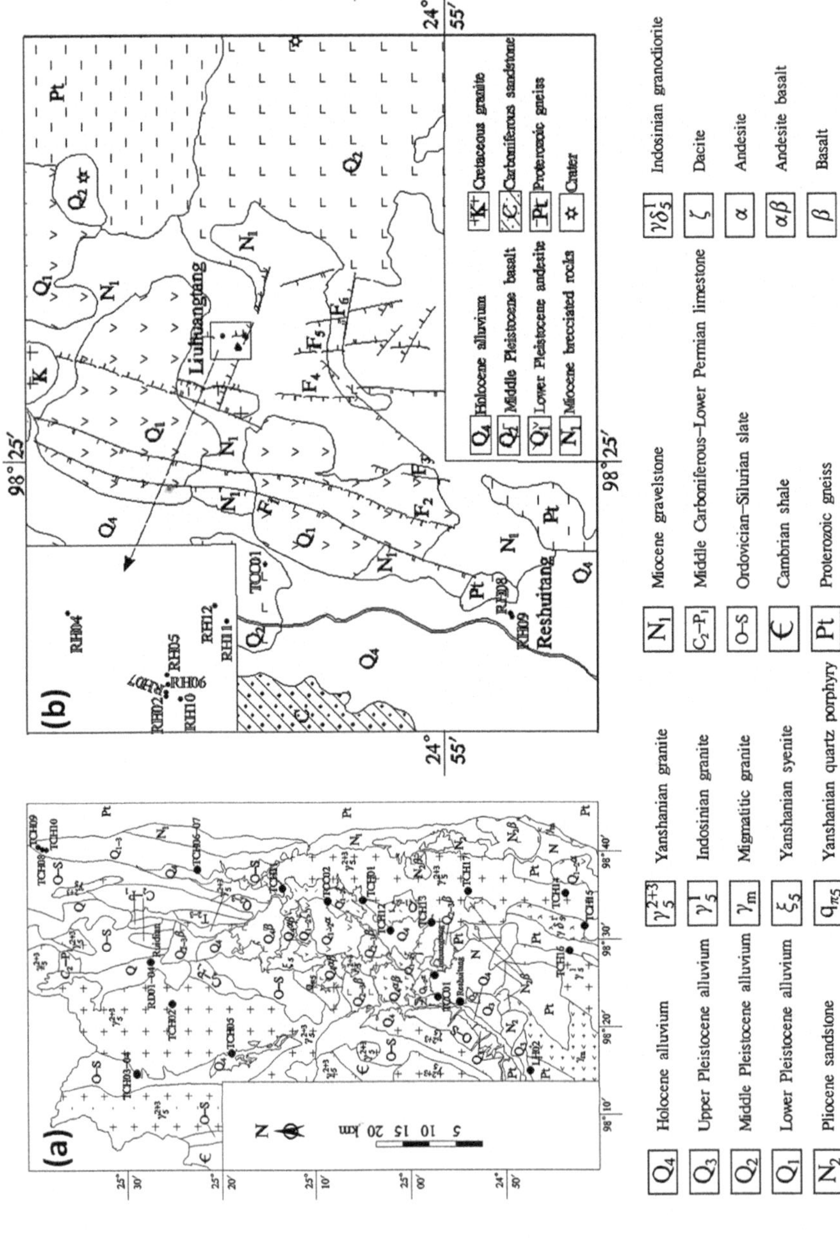

Fig. 3.13 Geological maps of (a) Tengchong and (b) Rehai (reproduced with permission from [35])

Table 3.5 Natural gas composition of Well Wulong 1 (adapted and translate from [21])

Depth (m)	He (%)	H_2 (%)	N_2 (%)	CO_2 (%)	C_1 (%)	C_2 (%)	O_2 (%)	Ar (%)	CO (%)
3140–4620	0.04	0.03	3.74	96.19	Small	0	0	–	–
	0.04	0.07	1.85	98.04	Small	0	0	–	–
	0.04	0.07	1.67	98.22	Small	0	0	–	–
	0.044	0.019	2.052	96.98	0.070	–	0.21	0.024	0.13
	0.075	0.156	4.862	94.86	0.046	–	–	–	–
2411.3–2425	0	33.14	53.54	7.08	1.34	0.02	–	–	–
	–	18.33	44.38	4.66	1.59	0.19	0.30	0.10	30.43
	0.182	43.794	50.549	3.526	1.883	0.016	–	0.018	–
	0.152	40.881	51.195	6.040	1.671	0.016	–	0.026	–

3.3 Australia

Australia's natural hydrogen are predominantly situated within sedimentary basins, including the Cooper, Gippsland, Otway, Southeast Queensland, Darling, and multiple basins across Western Australia (Fig. 3.16), in addition, several prospective areas potentially hosting natural hydrogen have been identified (Fig. 3.17). These regions host geological frameworks conducive to H_2 generation and retention, often capitalizing on pre-existing hydrocarbon infrastructure. Additionally, ongoing research efforts targeting H_2 production technologies further enhance the viability of these systems. Historically, a limited number of documented occurrences of hydrogen-enriched natural gases (up to 0.1 mol% H_2) have been reported from Australian reservoirs (Fig. 3.16) [5].

3.3.1 Western Australia

The majority of natural hydrogen discoveries are associated with two key geological characteristics: circular to sub-circular surface depressions and the presence of Fe-rich Precambrian basement rocks, where redox reactions may produce H_2 (Fig. 3.18). In Western Australia, these conditions are met within the Archean Yilgarn Craton, which contains iron-enriched metamorphic and igneous rocks conducive to hydrogen generation, as a result, numerous hydrogen occurrences have been discovered in this area, followed by drilling exploration (Fig. 3.19). Although direct evidence of free hydrogen is lacking, circumstantial indicators, such as aligned depression fields, vegetation anomalies, and surface discoloration, suggest subsurface hydrogen migration and potential seepage [38].

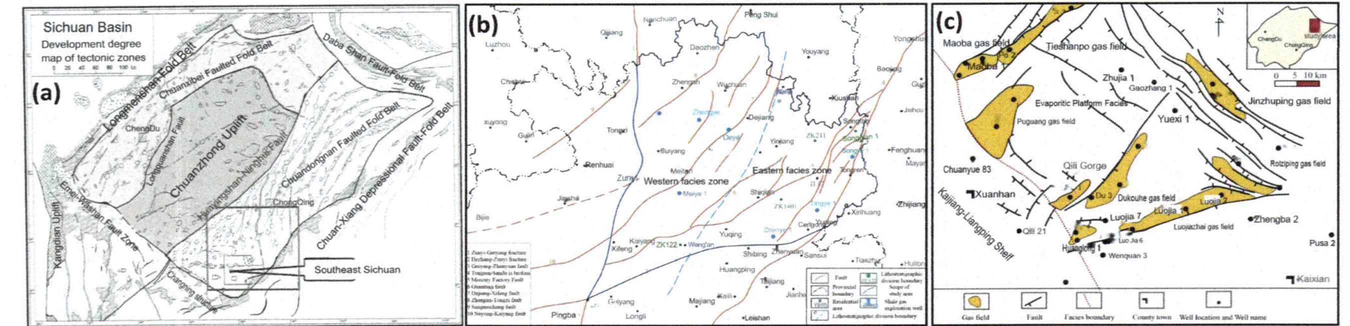

Fig. 3.14 The geological structural framework and hydrocarbon distribution in the Sichuan Basin and Its Periphery. **a** Map of the development degree of tectonic zoning in the Sichuan Basin. **b** Schematic diagram of fault faults in northern Guizhou area and lithofacies zoning map of Niutitang Formation. **c** The distribution of high-sulfur gas fields in the Feixianguan Formation of northeastern Sichuan and the comprehensive stratigraphic column of the eastern Sichuan region (modified from [15, 16, 22])

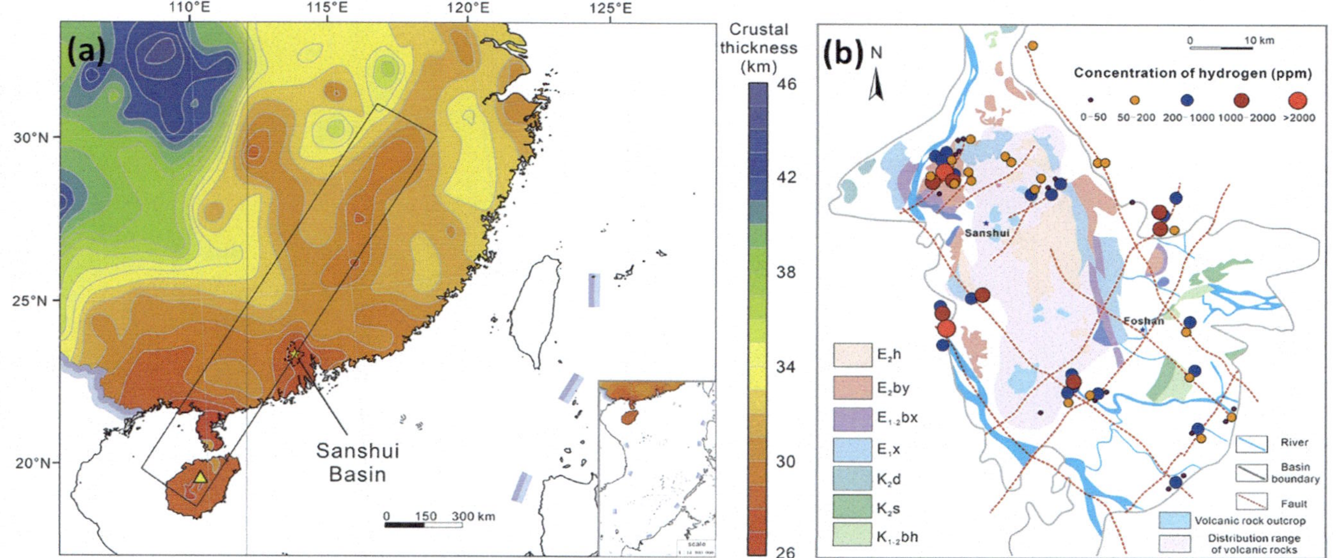

Fig. 3.15 Natural hydrogen resource in Sanshui Basin, South China (reproduced with permission from [11]. **a** The contour map of the crustal thickness of South China. The yellow triangle denotes the location of the Hainan mantle plume, while the Sanshui Basin is indicated by yellow star (modified from [36]). **b** Distribution of surface hydrogen seepage sites and corresponding hydrogen concentrations within the Sanshui Basin (modified from [37])

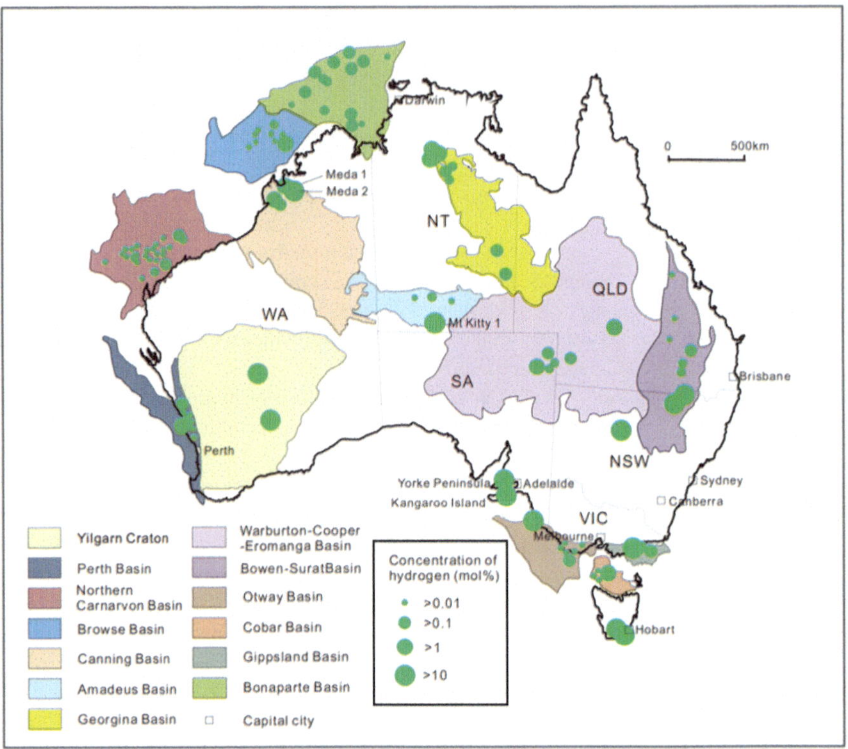

Fig. 3.16 A geographic map summarizing the measurable concentration of hydrogen resources (in mol%) in Australia (adapted from [5])

3.3.1.1 Canning Basin

Drill pipe tests conducted in the 1950s at Meda#1 (Fig. 3.20) at a depth of 2,685 m in the Canning Basin of Washington, Australia contained more than 95% H_2 (Fig. 3.21), in addition to natural hydrogen found in Well Mt Kitty 1 (11% H_2) at a depth of 2,295 m in the Amadeus Basin, Northern Territory, Australia (Fig. 3.22) [38, 40].

3.3.1.2 Northern Perth Basin

The circular depression, also known as the "fairy circle", represents the most obvious manifestation of natural hydrogen flow. Circular depressions appear as thicker sedimentary sequences in relatively young basins, such as the Perth Basin in Washington State, Australia (Fig. 3.23a). And giant archaic cratons, such as the Yilgarn craton in Washington, Australia, have little or no regolith cover (Fig. 3.23b) [38].

The Northern Perth Basin, situated west of the Yilgarn Craton, is a structurally intricate basin characterized by its eastern margin being governed by the Darling Fault (Fig. 3.24). A total of seventy-nine soil gas samples were collected between Moora and Pingarrega

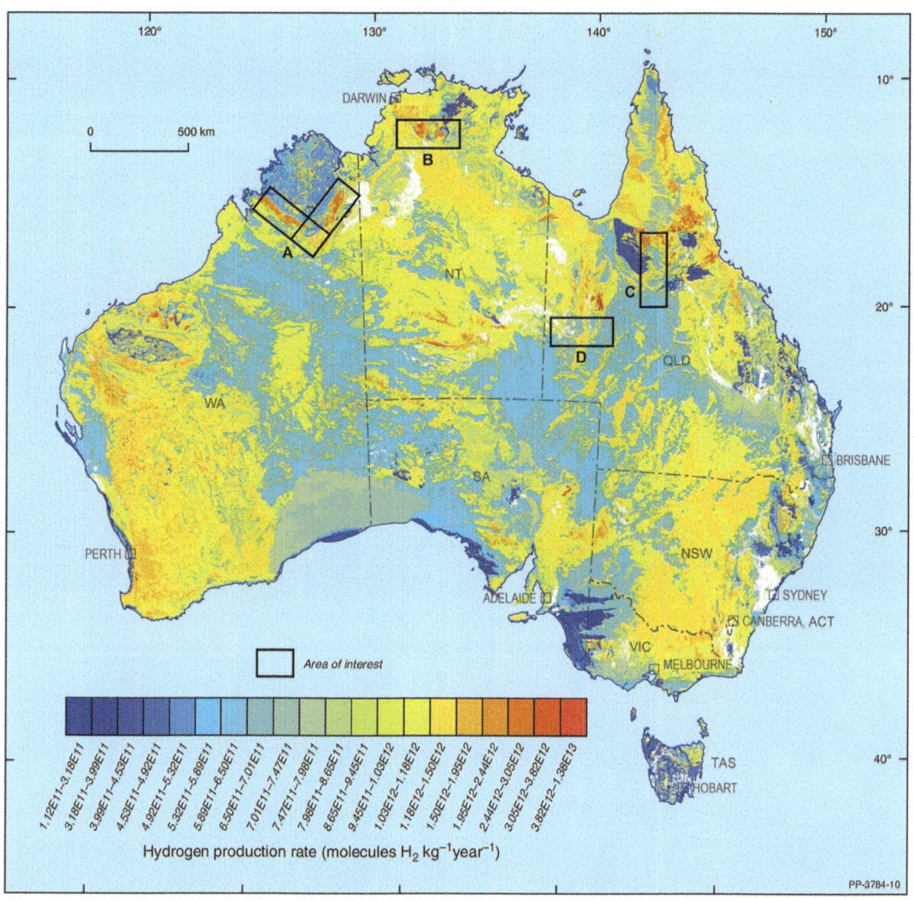

Fig. 3.17 The radiolytic hydrogen generation rate (in molecules H_2 kg^{-1} $year^{-1}$) across onshore Australia highlights several regions of interest. (A) The King Leopold and Halls Creek Orogens exhibit anomalous H_2 production rates, justifying further exploration to assess potential trap structures. (B) Similarly, the Pine Creek Orogen shows elevated radiolytic yields that merit additional study for reservoir potential. (C) The region west of the Georgetown Inlier, beneath the Carpentaria Basin, is also identified as a promising target for hydrogen accumulation. (D) Likewise, the area south of the Mt Isa Inlier, underlying the Georgina Basin, represents another prospective zone for hydrogen exploration (reproduced with permission from [5])

(Fig. 3.25a), revealing H_2 concentrations spanning from 0 to 96 ppm, with the maximum concentration detected in proximity to the Darling Fault. Regional measurements near Moora 3 (Fig. 3.25b) demonstrated a clear correlation between near-surface lithology and H_2 anomalies. To clarify the hydrogen system in the study area, a conceptual model was constructed, synthesizing proposed mechanisms for gas generation, migration pathways, as well as possible entrapment and leakage processes (Fig. 3.26) [43].

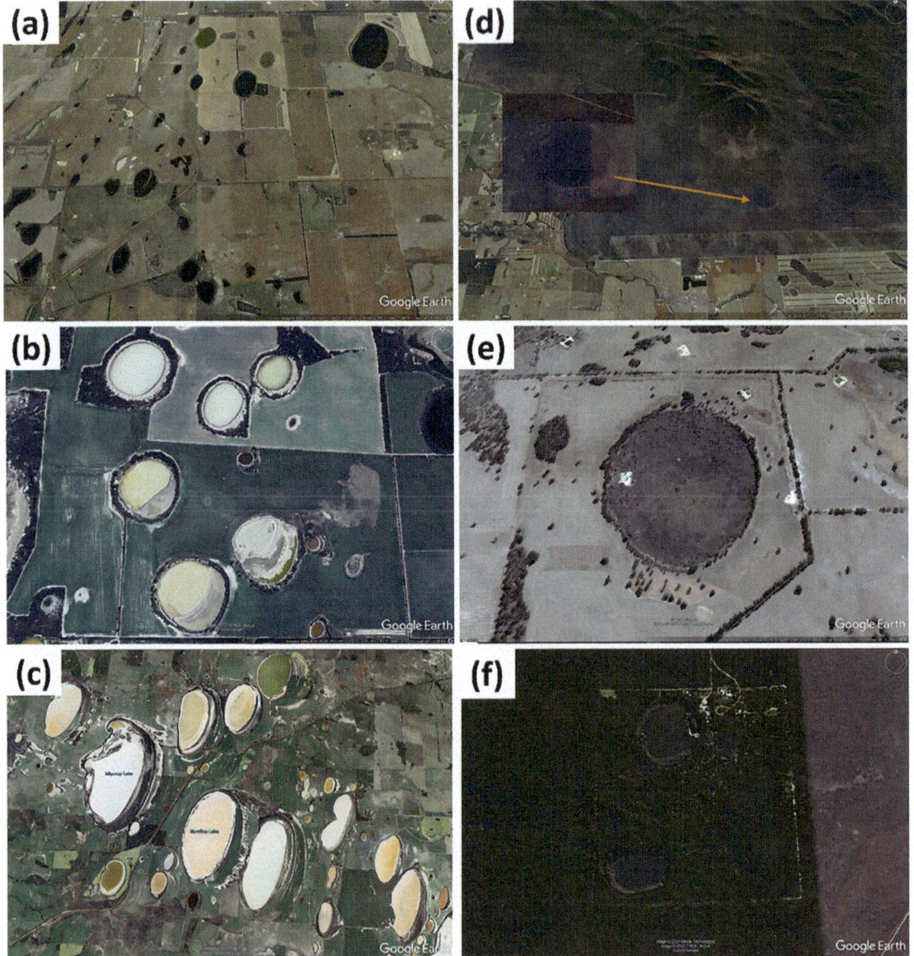

Fig. 3.18 Examples of circular to subcircular depressions in Western Australia. Several features exhibit no clear association with drainage networks (e.g., **a**, **b**), while others are aligned with paleochannels (e.g., **c**). Certain depressions occur near foothills (e.g., **d**; inset provides detailed view), some are desiccated (e.g., **e**), and others are partially or fully vegetated (e.g., **f**) (modified from [38])

Figure 3.27 presents a meticulously reconstructed 3D visualization of the Perth Basin, illustrating the spatial relationships among its principal tectonic units. To the east of the Darling Fault, a series of granite and greenstone belts (Fig. 3.28) develop and are formed on the west side by sedimentary sequences from the Perth Basin, where there are clusters of circular depressions with the following H_2 concentrations in the topsoil layer (< 1 m) (Fig. 3.29) [44].

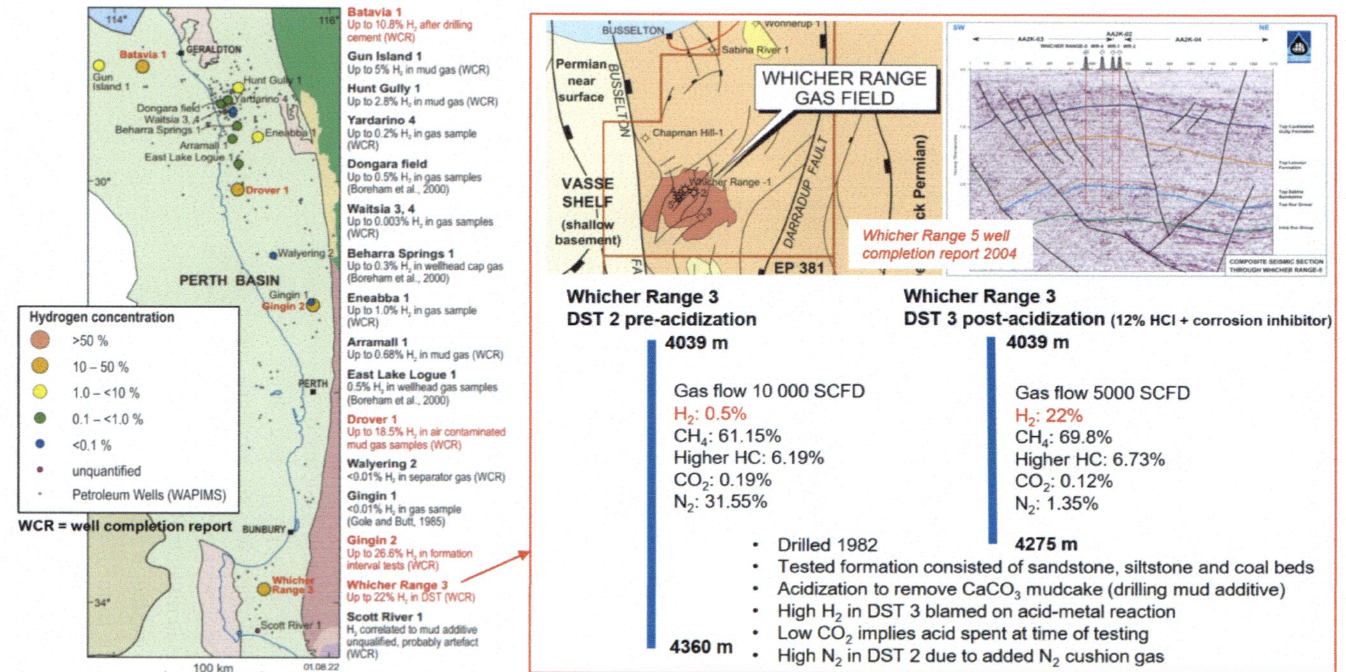

Fig. 3.19 Geological map of Perth Basin containing the natural hydrogen occurrences and exploration well locations in Western Australia (source from: Department of Mines, Industry Regulation and Safety, Western Australia, Australia [41])

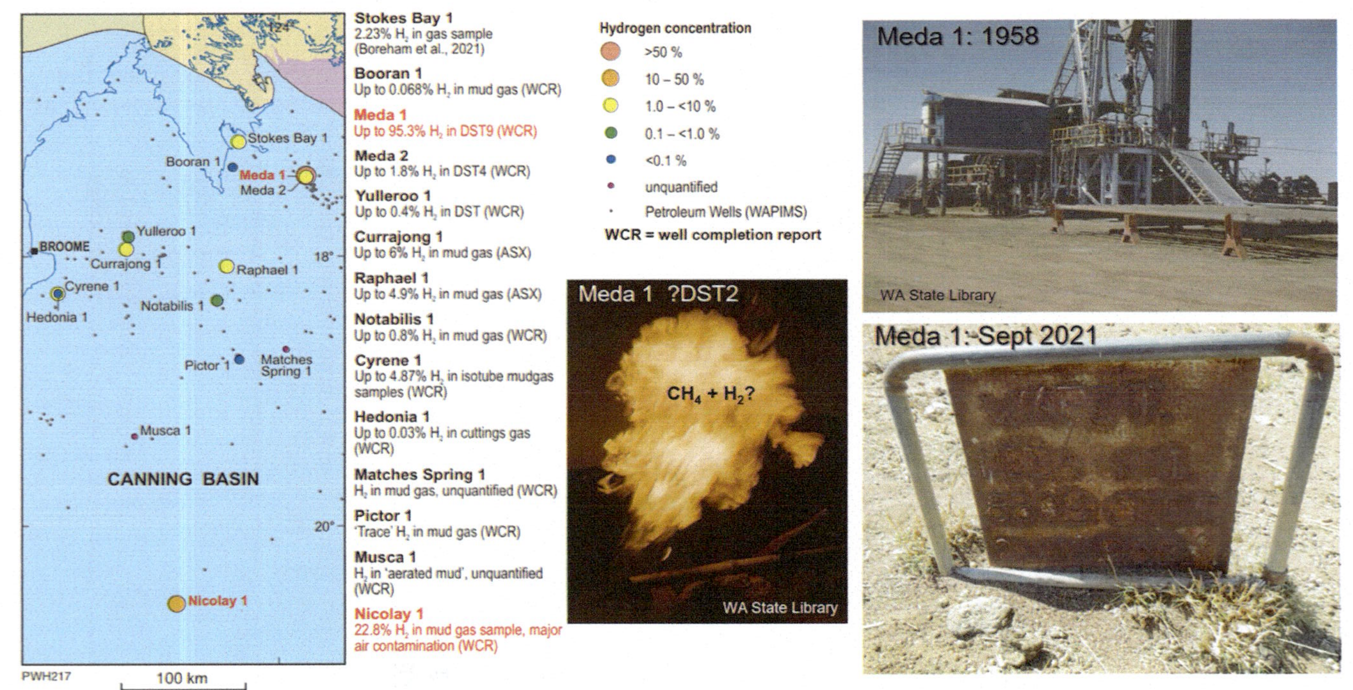

Fig. 3.20 Details of Meda-1 well in Western Australia (source from: Department of Mines, Industry Regulation and Safety, Western Australia, Australia [41])

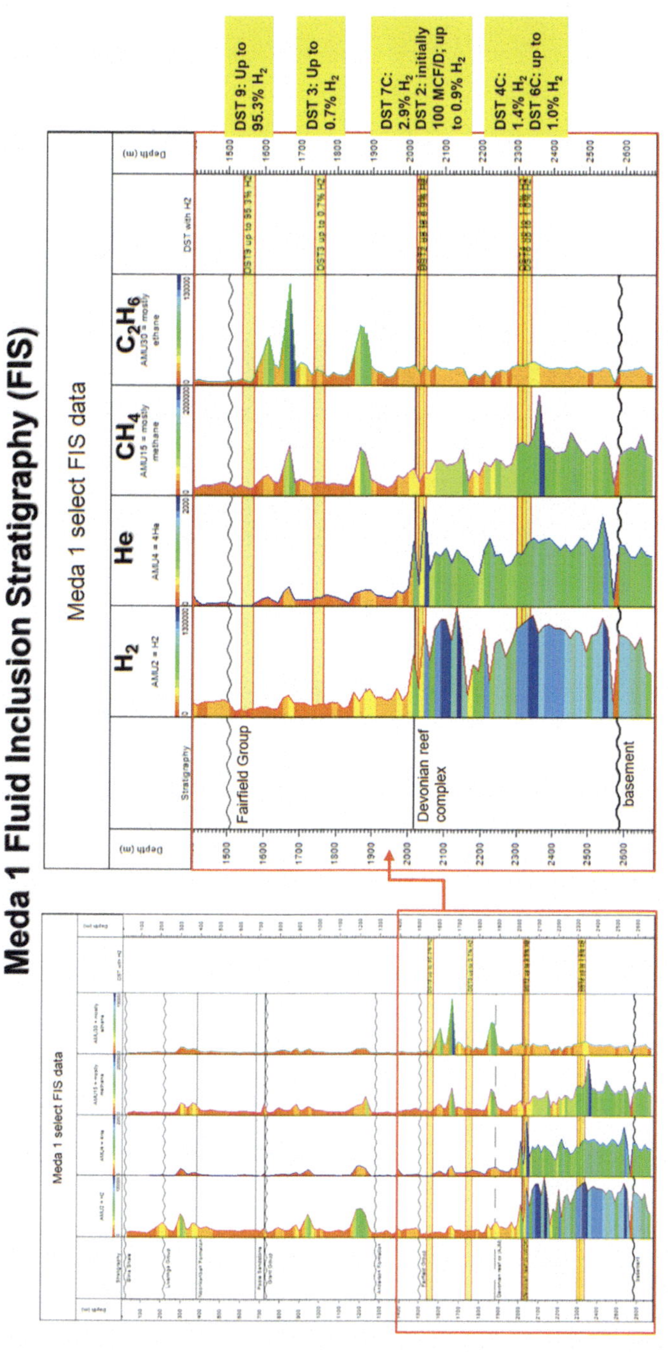

Fig. 3.20 (continued)

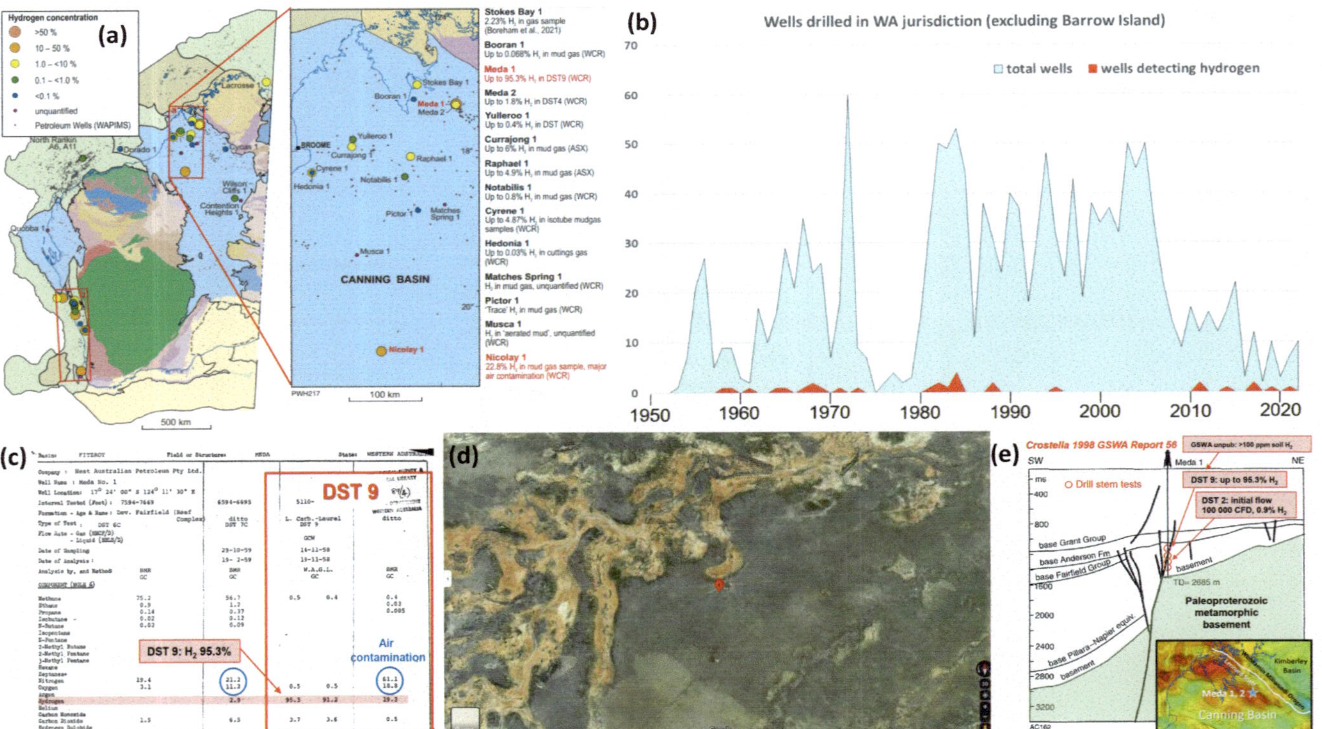

Fig. 3.21 Meda-1 Well in Western Australia (source from: Department of Mines, Industry Regulation and Safety, Western Australia, Australia [41]). **a** Hydrogen occurrences in Western Australian petroleum wells and fields are located. **b** Wells drilled in WA jurisdiction (excluding Barrow Island). **c** Meda-1 well shows up to 95.3% hydrogen content. **d** Meda-1 well location (source from: Department of Mines, Industry Regulation and Safety, Western Australia, Australia). **e** Meda-1 well section (modified from [40, 41])

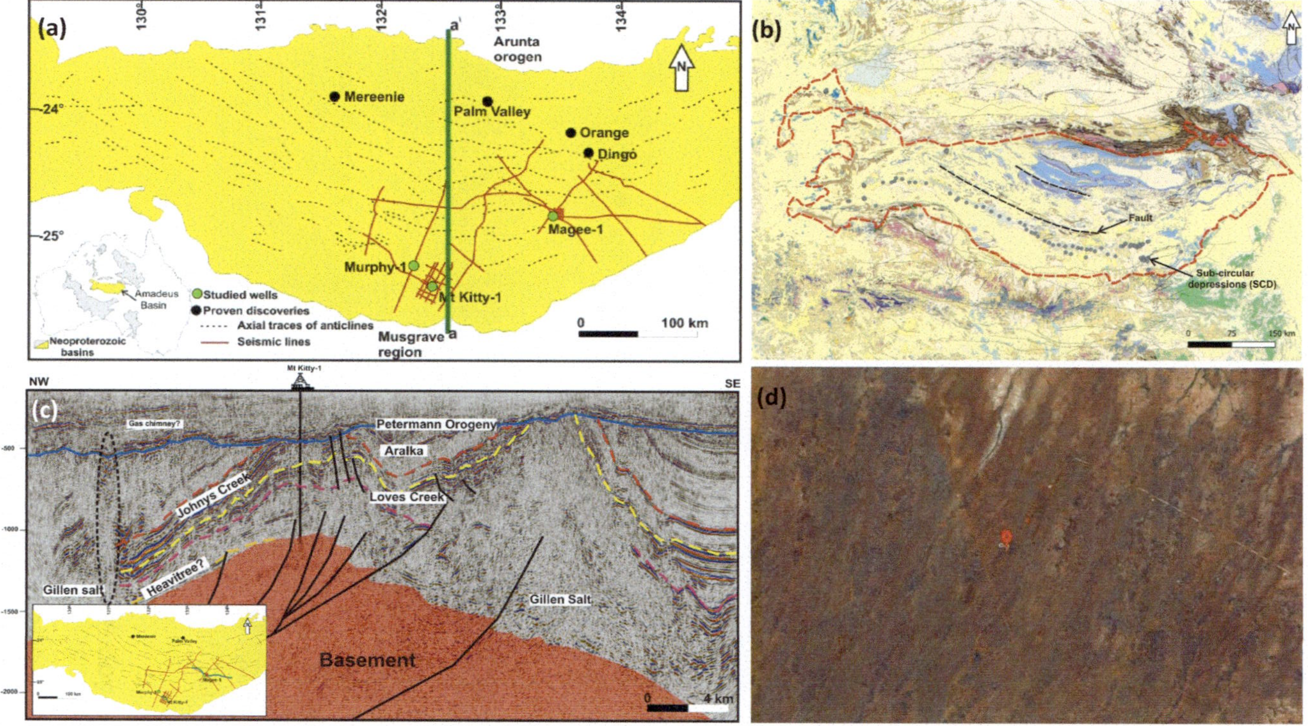

Fig. 3.22 Mt Kitty 1 Well in Western Australia. **a** Location map depicting the investigated wells and confirmed gas discoveries within the Amadeus Basin, central Australia. **b** Map showing the distribution of sub-circular depressions (SCDs), represented as grey dots, in the southern Amadeus Basin. These depressions align parallel to the dominant NW–SE oriented fault systems traversing the basin. A notably higher density of SCDs is observed eastward, adjacent to wells that have confirmed the presence of hydrogen. **c** Interpreted 2D seismic profile showing the principal fault systems within the Mt Kitty structure. **d** Google Maps view of the Mt Kitty-1 well location (Northern Territories, Australia), with coordinates sourced from well records (modified from [40, 42])

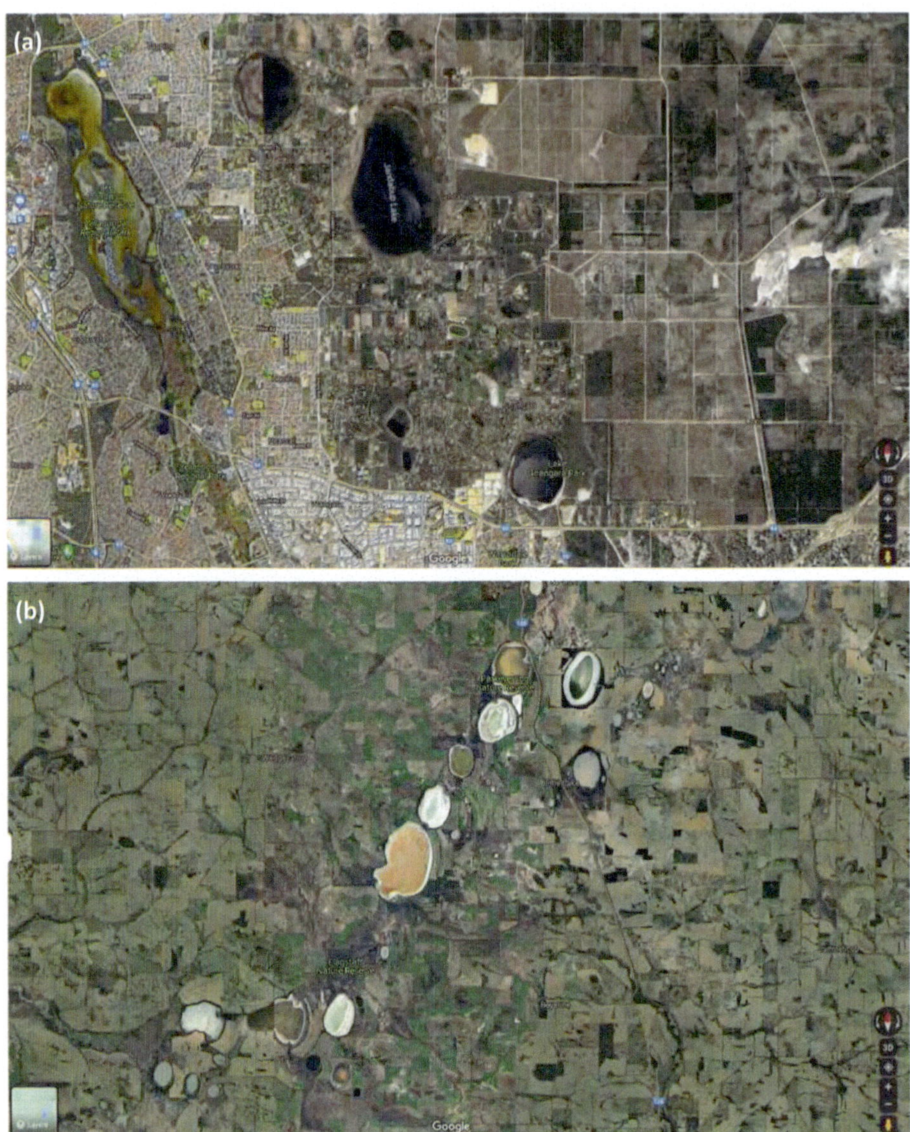

Fig. 3.23 Examples of Terminal Lakes in the Differing Geologic Settings of the Perth Basin and Yilgarn Craton. **a** Joondalup Lake and surroundings in Perth Basin, Western Australia, Australia. **b** The region encompassing Lake Norring and its adjacent areas, as identified within the Yilgarn Craton of Western Australia (reproduced with permission from [40])

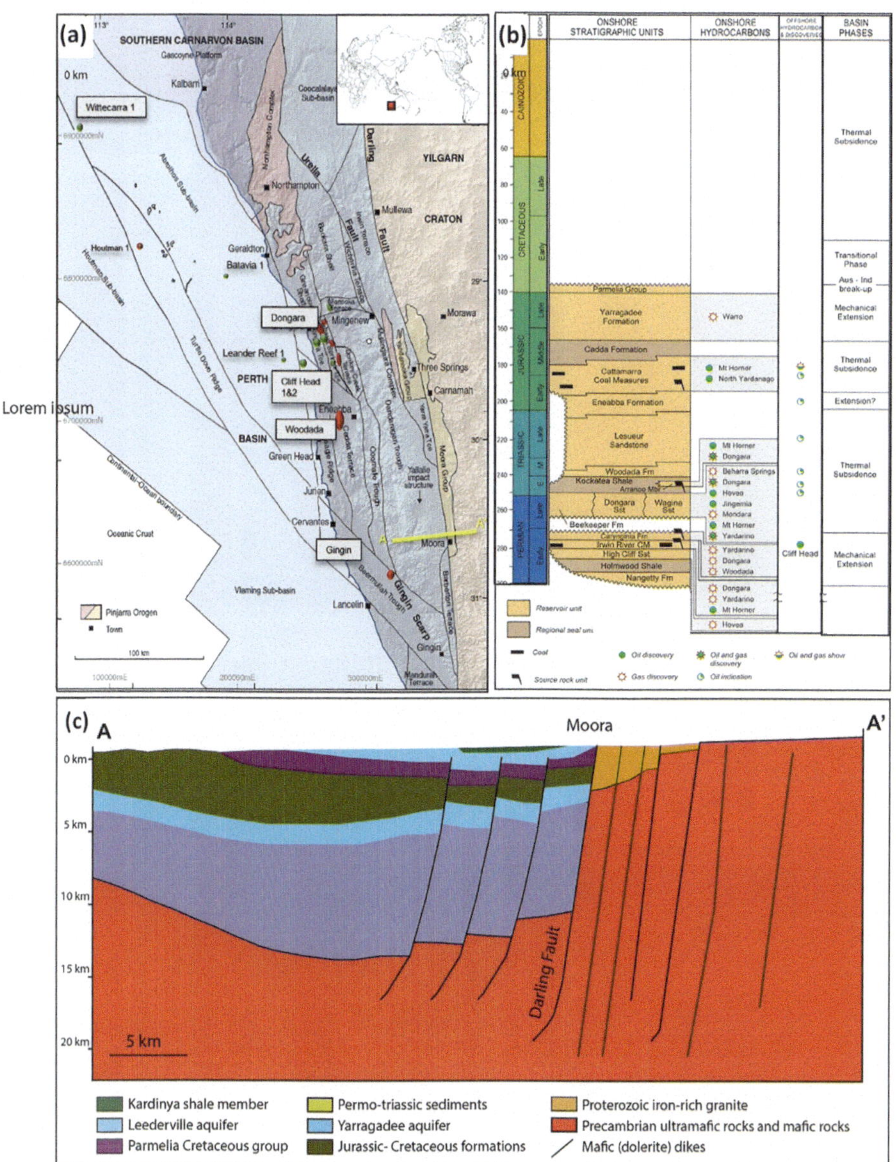

Lorem ipsum

Fig. 3.24 North Perth Basin geological settings. **a**, **b** Structural map illustrating hydrocarbon discoveries; stratigraphic column, petroleum system components, and key tectonic events of the North Perth Basin. **c** Interpreted geological and hydrogeological cross-section looking westward through the Moora area, with location indicated in (**a**) (reproduced with permission from [43])

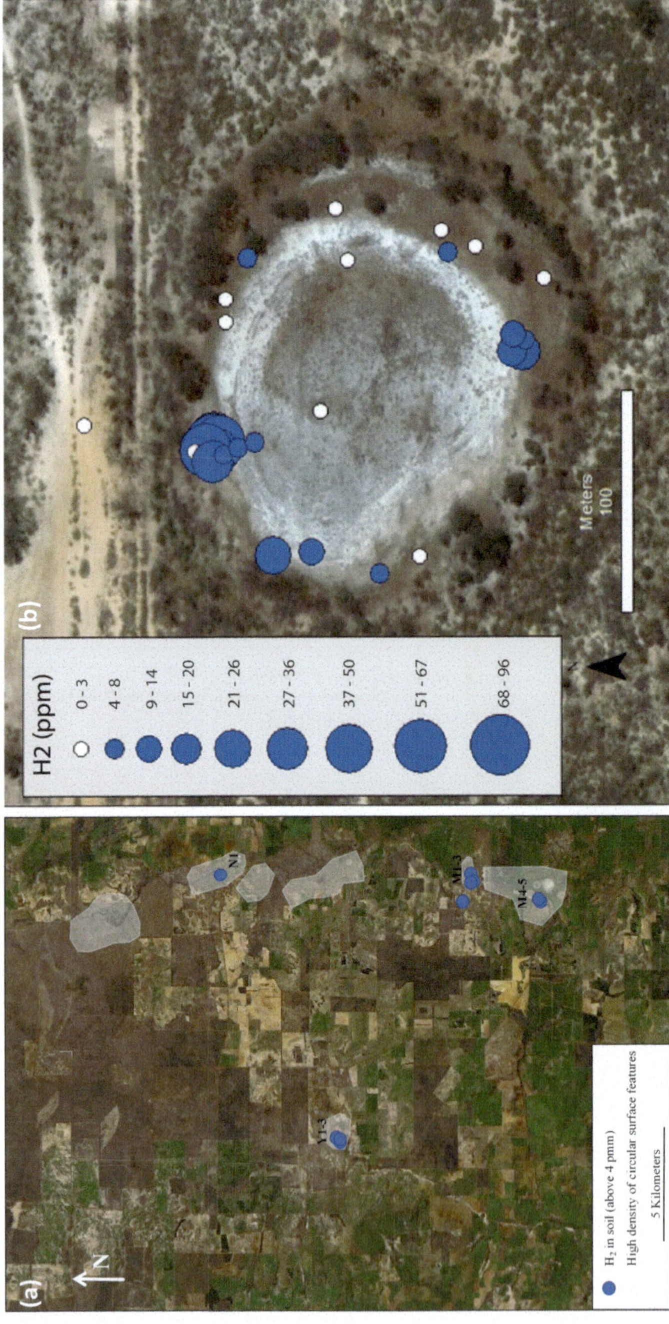

Fig. 3.25 Hydrogen soil-gas measurements and detailed characterization of a circular depression in the Moora-Pingarrega area. **a** Hydrogen soil-gas survey conducted in the Moora–Pingarrega region, overlain on an Esri World Imagery base map [45]. **b** Detailed hydrogen soil-gas analysis at the Moora 3 circular depression site, displayed on an Esri World Imagery base map [43]

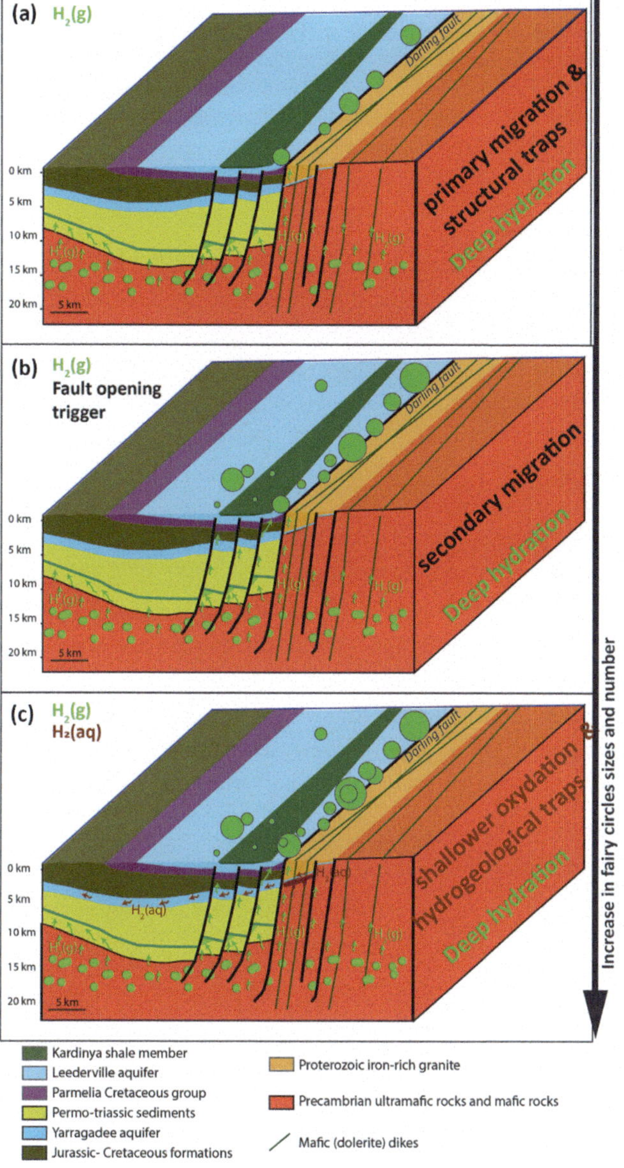
(a) H2(g)
Darling fault
primary migration & structural traps
Deep hydration
0 km
5 km
10 km
15 km
20 km
5 km
(b) H2(g)
Fault opening trigger
secondary migration
Deep hydration
(c) H2(g)
H2(aq)
shallower oxydation & hydrogeological traps
Deep hydration
Increase in fairy circles sizes and number
Kardinya shale member
Leederville aquifer
Parmelia Cretaceous group
Permo-triassic sediments
Yarragadee aquifer
Jurassic- Cretaceous formations
Proterozoic iron-rich granite
Precambrian ultramafic rocks and mafic rocks
Mafic (dolerite) dikes

◀**Fig. 3.26** Conceptual models of the Moora-Pingarrega hydrogen system, integrating geological and hydrogeological components (see Fig. 3.24c). **a** H_2 generation via serpentinization of deep-seated ultramafic rocks, accumulation within structural traps or limited surface seepage. **b** Fault reactivation enabling vertical migration and secondary H_2 transport toward the surface. **c** Low-temperature alteration of mafic dikes and iron-rich lithologies, followed by aqueous hydrogen mobility within the aquifer system (reproduced with permission from [43])

3.3.1.3 Yilgarn Craton

The majority of natural hydrogen occurrences are typically linked to deposits exhibiting circular to subcircular morphological features, frequently situated within Precambrian terranes abundant in iron-rich metamorphic and igneous lithologies. These geological environments provide favorable conditions for H_2 generation through geochemical processes such as Fe^{2+} oxidation and H_2O reduction. The Yilgarn Craton in Western Australia meets both geological prerequisites (Fig. 3.30), and hydrogen seepage has been detected at the FF4 Site (Fig. 3.31) and The Grass Patch area (Fig. 3.32) within the craton, demonstrating this spatial correlation [43].

The Yilgarn Craton comprises predominantly metamorphosed granitic, volcanic, and sedimentary lithologies. According to Previous studies [47], the craton's tectonic evolution has been categorized into six distinct stratigraphic subdivisions (Fig. 3.30). The region's extensive iron-rich lithologies and mafic–ultramafic assemblages position it as a potential source rock for H_2 generation. Furthermore, its intricate network of fault systems may facilitate H_2 migration. Eocene-aged sedimentary sequences and banded iron formations (BIFS) are interpreted to serve as effective reservoir units and impermeable caprocks, providing both storage capacity while inhibiting vertical H_2 migration [38].

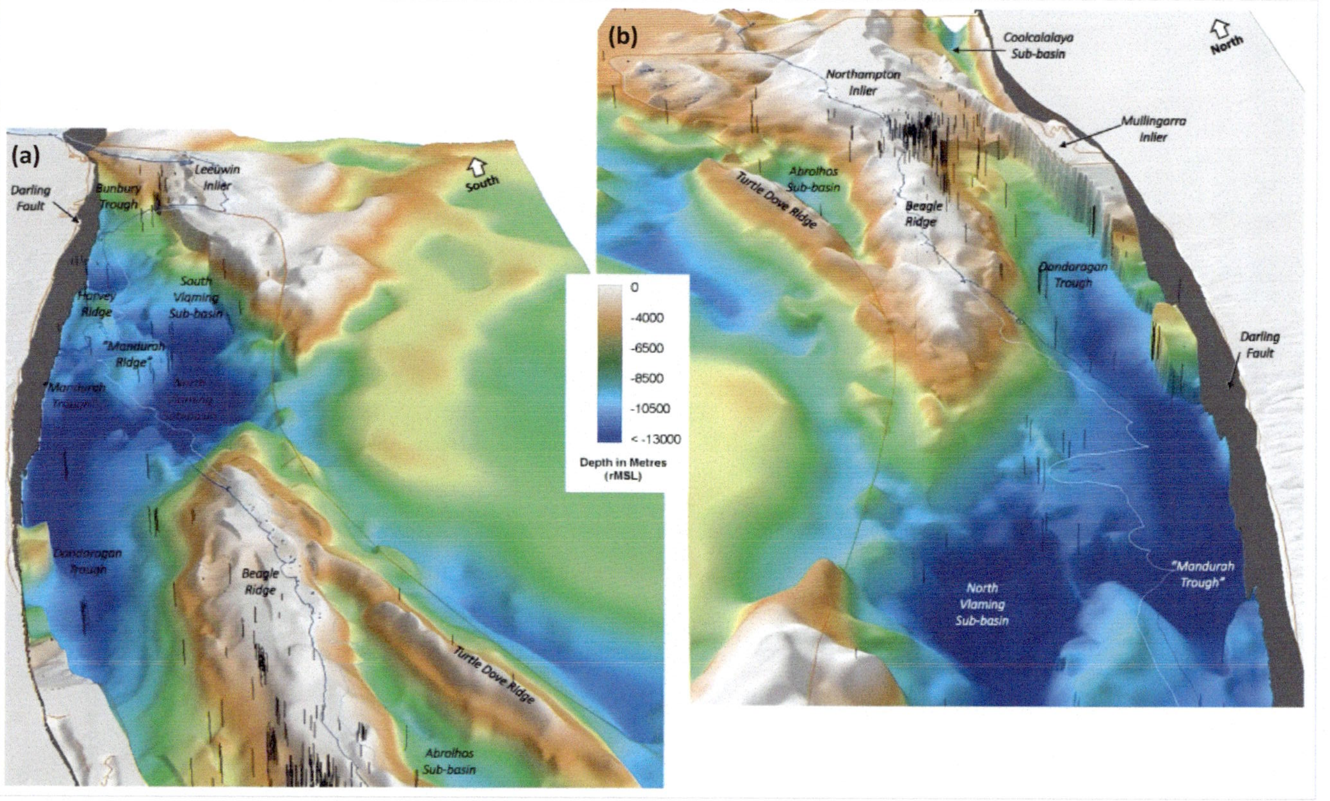

Fig. 3.27 Oblique three-dimensional perspective diagrams of the Perth Basin. **a** A northward-looking viewpoint. **b** A southward-looking viewpoint. Adapted from Geognostics Australia Pty Ltd. [44]

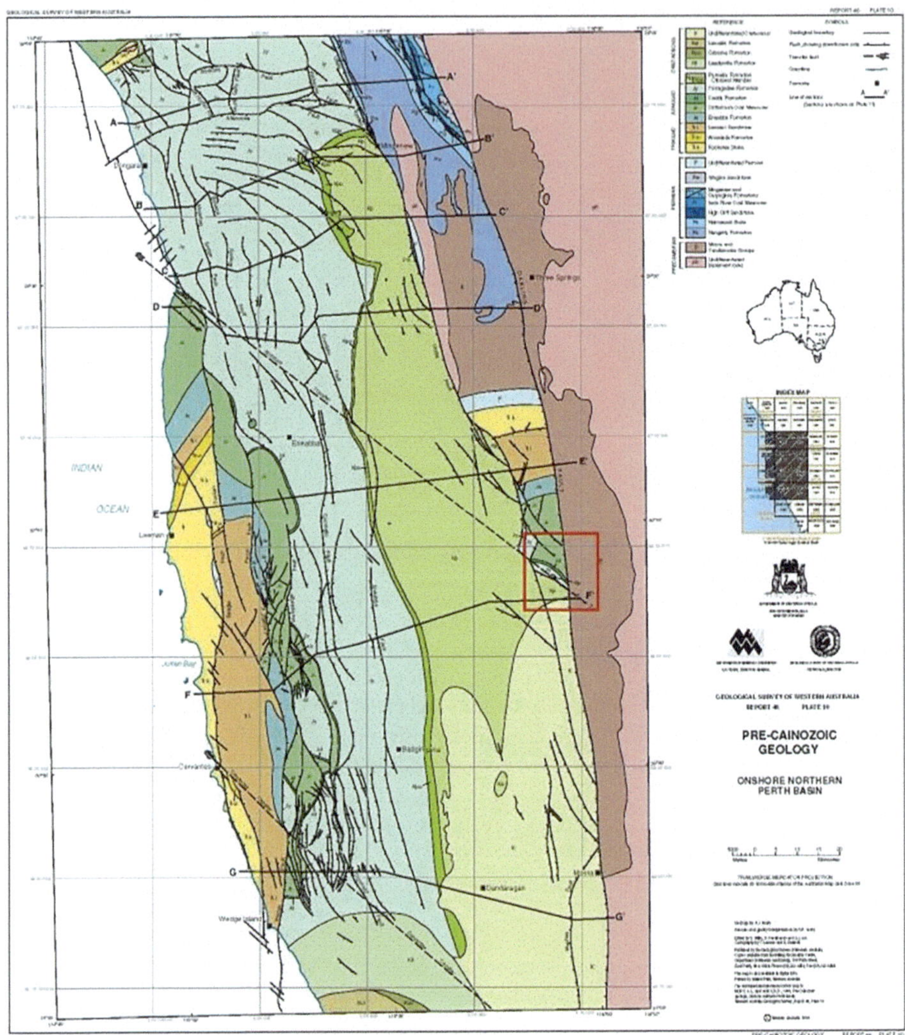

Fig. 3.28 Structural map of the North Perth Basin, depicting pre-Cainozoic geology [46]. The AOI is delineated by a red rectangle (reproduced with permission from [44])

The Grass Patch area is located on the southeastern margin of the Yilgarn Craton. Eleven hydrogeochemical structures were sampled across this region, as illustrated in Fig. 3.32a. Recorded concentrations spanning 0–48 ppm, with the highest H_2 values peaking at the boundary of the adjacent nature reserve (Fig. 3.32b). The lowest-concentration features correlate with elongated, "terrain-aligned" structures, which were preliminarily

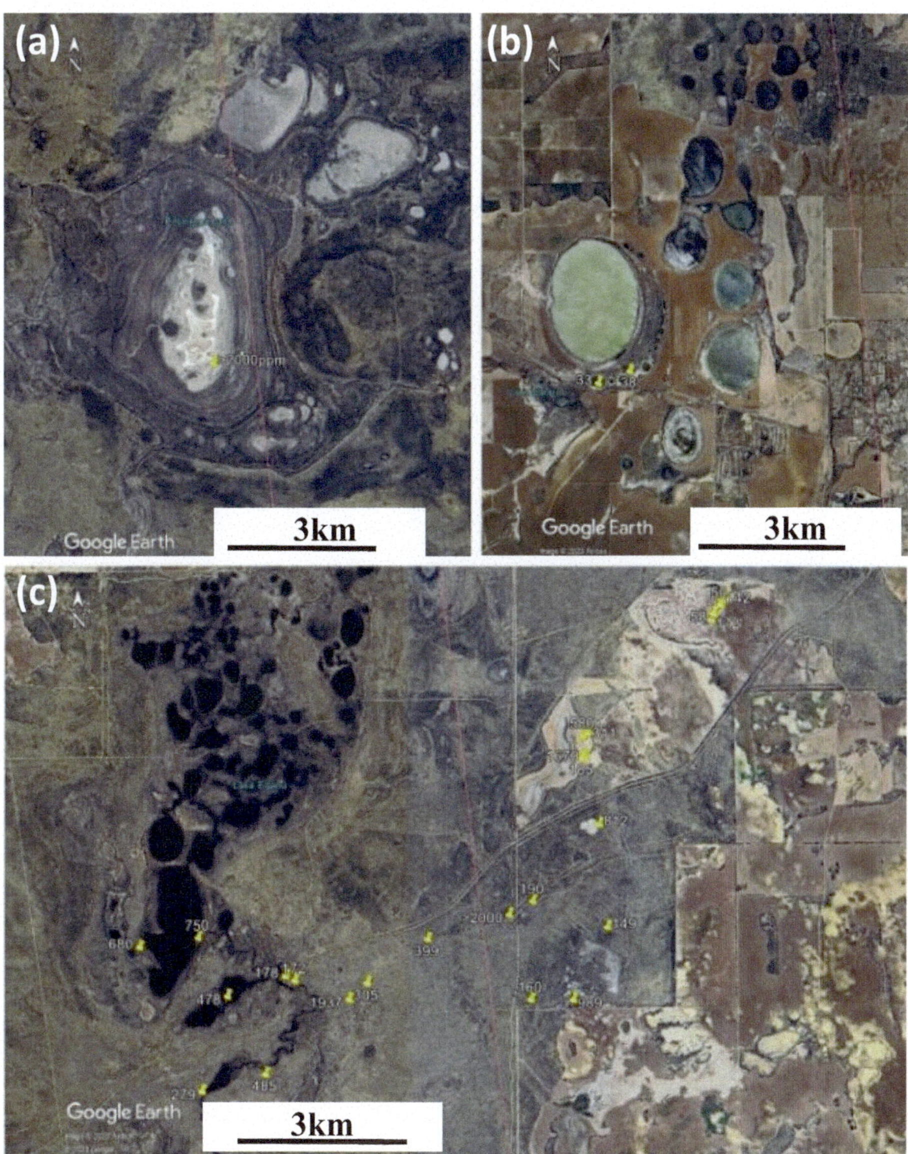

Fig. 3.29 Circular depressions within the AOI were investigated, including **a** Lake Pinjarrega, **b** Lake Dalaroo, and **c** Lake Eganu. Hydrogen concentrations (in ppm) are indicated by yellow pins, and the Darling Fault is represented by a solid pink line (reproduced with permission from [44])

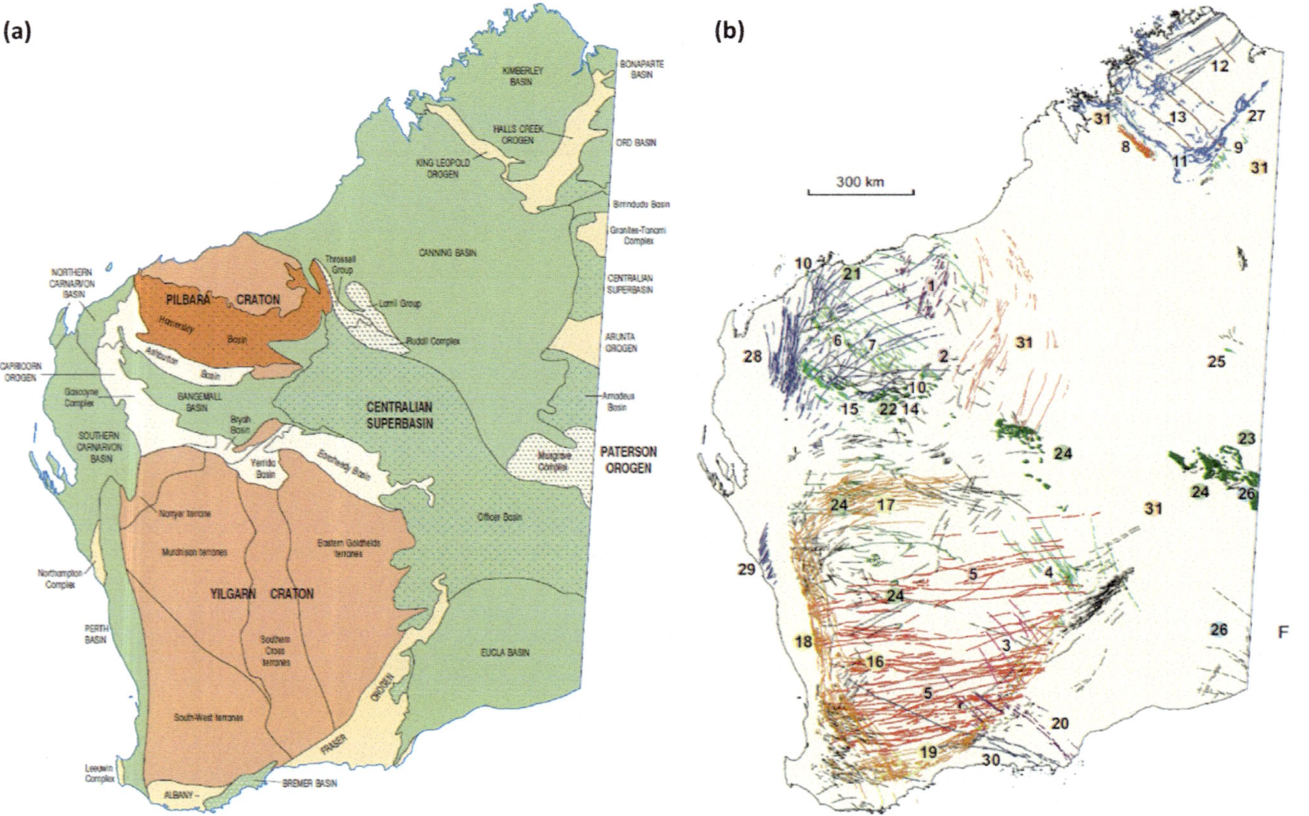

Fig. 3.30 Composite map of the Yilgarn Craton: Terrane Subdivision and distribution of Dyke Swarms. **a** Based on its tectonic evolution, the Yilgarn Craton in Western Australia is divided into six terranes [47]. **b** Distribution map of dyke and sill suites in Western Australia, with individual dyke sets color-coded for distinction (reproduced with permission from [38, 49])

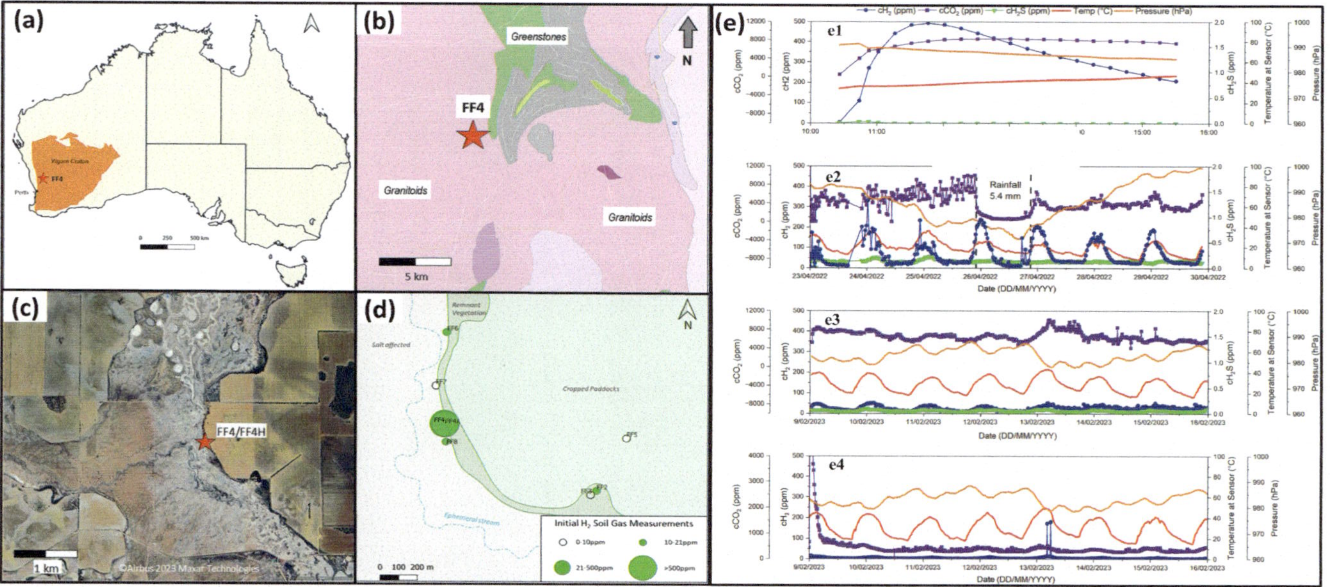

Fig. 3.31 Integrated characterization of a Natural Hydrogen Seep at the FF4 Site, Yilgarn Craton, Western Australia. **a** Sample location FF4 within the Yilgarn Craton. **b** Interpreted bedrock geology at the FF4 site. **c** Surface morphology featuring an ephemeral drainage channel and saline floodplains. Google Earth imagery courtesy of Maxar Technologies, modified. **d** Outcomes of a regional soil gas survey revealing anomalous hydrogen concentrations at the FF4 locality. **e** Hydrogen detection correlated with (a–d), displaying gas concentrations (H_2, H_2S, CO_2) alongside atmospheric pressure and temperature recorded at the sensor: **a** WHALI—20 February 2022, immediately post-drilling; **b** WHALI—23–30 April 2022; **c** WHALI—9–16 February 2023; **d** EVA—9–16 February 2023 (reproduced with permission from [50])

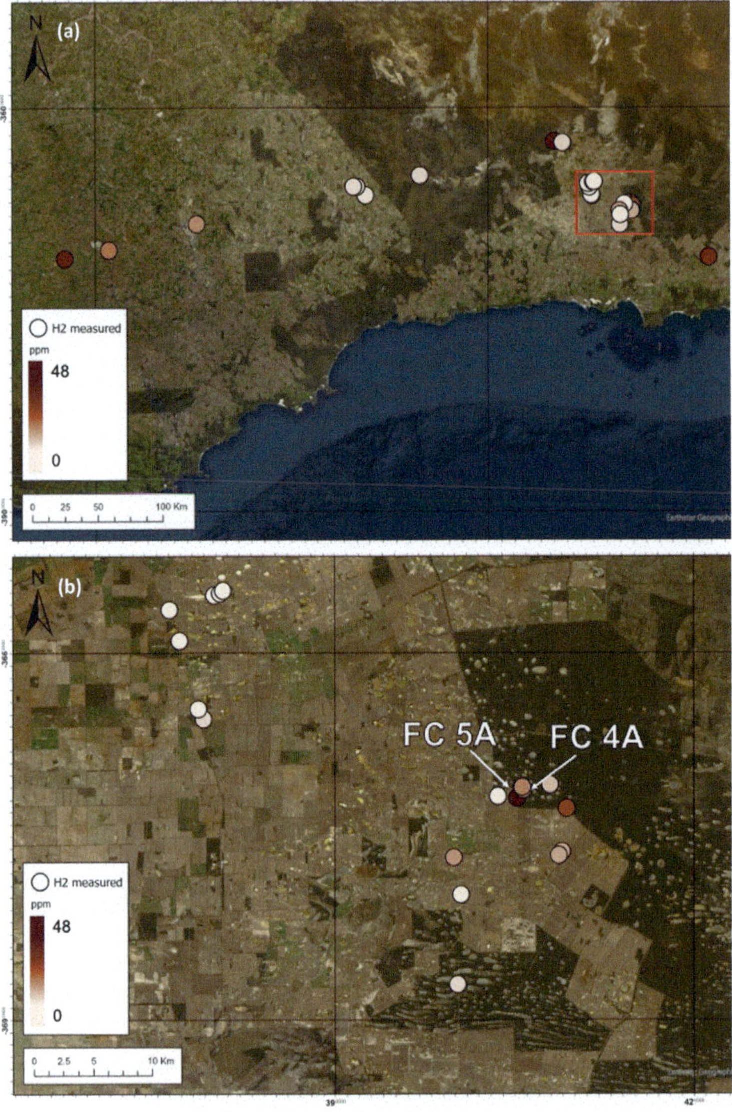

Fig. 3.32 Regional Context and Localized Hydrogen Measurements in Southern Western Australia. **a** Hydrogen concentration data obtained from multiple sampling campaigns across southern Western Australia. The red rectangle delineates the study area examined in this work. **b** Distribution of hydrogen measurements within the study area. FC-5A and FC-4A refer to monitored features. Data in the northwestern section were acquired during a separate field campaign (Imagery courtesy of Earthstar Geographics) (reproduced with permission from [48])

classified as non-hydropermetic during desktop assessments. H_2 concentrations demonstrate a distinct temporal trend: an initial rapid ascent is followed by a gradual yet sustained decline [48].

3.3.2 Southern Australia

3.3.2.1 Kangaroo Island and the York Peninsula

H_2-enriched natural gas has been documented in paleo-boreholes across South Australia's York Peninsula and Kangaroo Island [5], regions underlain by Precambrian basement rocks. The geological framework comprises Cambrian sedimentary sequences overlying a basement composed of Palaeozoic to Mesozoic felsic magmatic clasts (Fig. 3.33). Two primary H_2-generation mechanisms are identified in this setting: First, H_2 is generated through radiolytic decomposition of water within uranium-, thorium-, and potassium-bearing granites. Second, supplementary H_2 production occurs via water–rock interactions (WRIs) involving iron-rich felsic lithologies, exemplified by the Precambrian Roxby Downs granite [51].

About 84% H_2 was reported in a borehole at a depth of 508 m on Kangaroo Island, South Australia. The geology of Kangaroo Island is similar to that of Western Australia, with many circular depressions in the area [38].

The Minlaton oil drilling well (Fig. 3.34a) in the central York Peninsula, South Australia, was located approximately 10 kms east of the town of Minlaton in Ramsay section 112 of the map (Fig. 3.34b), and multiple H_2 outgassing was observed during drilling [5].

3.3.2.2 Cooper Basin: Queensland and South Australia

Geological Features: This sedimentary basin is globally recognized for its abundant hydrocarbon resources. The Cooper Basin predominantly comprises shale and sandstone formations, which exhibit potential for H_2 retention (Fig. 3.35). Hydrogen Potential: H_2 generation in the region could be facilitated through processes such as the water–gas shift reaction, leveraging existing gas reserves. In the Cooper Basin, 26 wells have gases with $H_2 > 0.1$ mol% and 9 of these wells having gases with $H_2 > 1.0$ mol%. H_2 in the Cooper Basin is mainly derived from the thermal decomposition of terrestrial plant-derived organic matter. Under conditions of elevated thermal maturity and temperature, the Patchawarra Formation source rock facilitated the generation of hydrogen (H_2) in the absence of methane [52].

3.3.2.3 Gippsland Basin: Location: Victoria's Offshore Area

Renowned for its abundant oil and gas reserves, the Gippsland Basin hosts significant geological structures exhibiting the potential capacity to support the entrapment and storage of H_2 (Fig. 3.36).

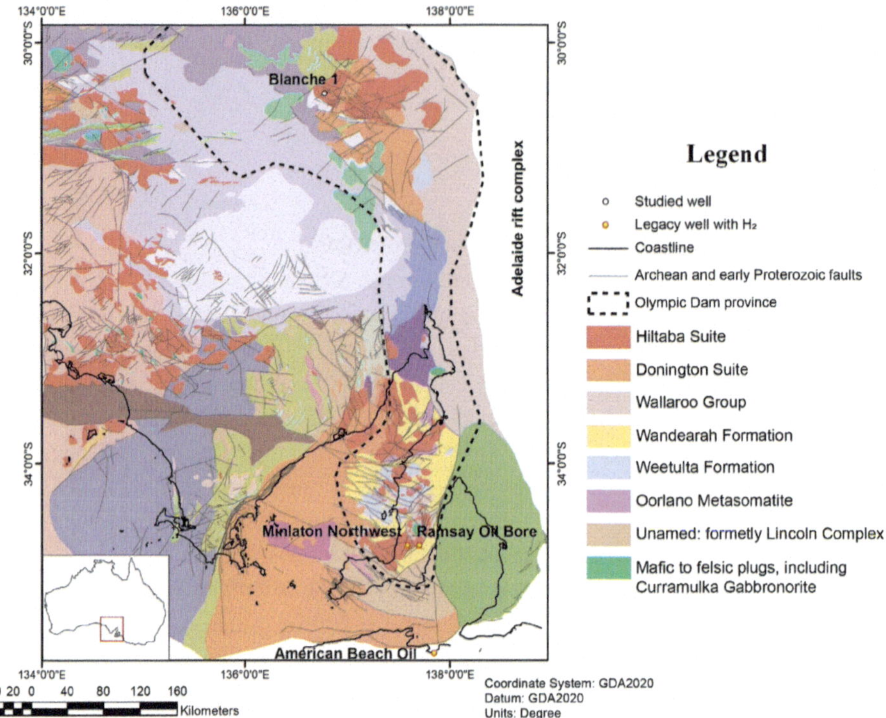

Fig. 3.33 Geological mapping illustrates the principal Archaean to Early Mesoproterozoic rock units across South Australia (SARIG database). The Blanche 1 well is situated within the northern region of the Olympic Dam Province. Three legacy wells, which yielded reservoir gas samples predominantly composed of H_2, are located in the southern sector of the same province. For lithological units not displayed in the legend, further details are available through the SARIG website (reproduced with permission from [51])

3.4 Africa: Mali Bourakebougou

Bourakebougou, located in Mali, West Africa, has emerged as a pivotal site for natural hydrogen extraction. The reservoir was identified serendipitously in 1987 during water well drilling, garnering substantial scientific interest because of its exceptionally high H_2 concentrations. Although the study of natural hydrogen systems remains a relatively nascent field, this site exemplifies significant potential for future energy applications. Figure 3.37 illustrates the field's large-scale geological framework. Hydroma Inc. has played a critical role in spearheading exploration and technological advancements at this location [54].

The Bourakebougou region hosts a well-developed multilayered reservoir system conducive to H_2 entrapment, attributed to oxygen-depleted and reducing environments

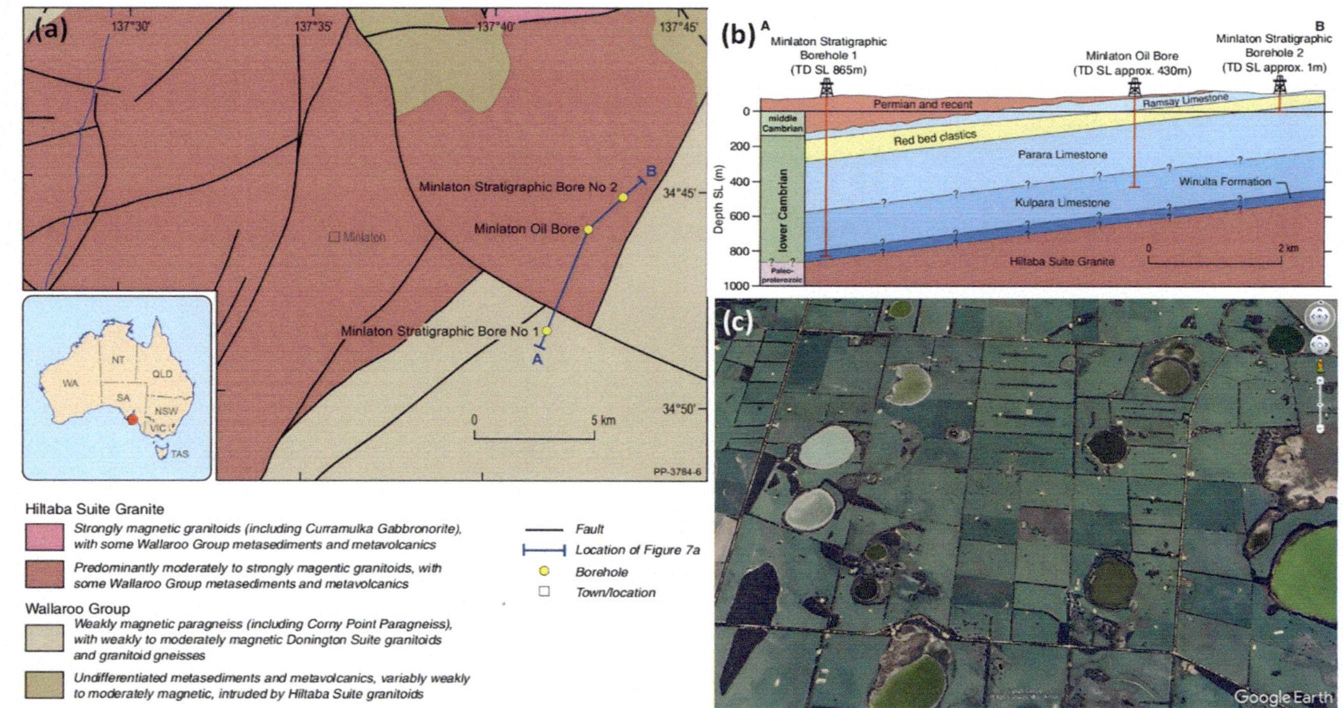

Fig. 3.34 The Gawler Craton-Stansbury Basin, South Australia. **a** interpreted basement geology in the vicinity of the Minlaton Oil Bore. **b** Stratigraphic well correlation among Minlaton Stratigraphic 1, Minlaton Oil Bore, and Minlaton Stratigraphic 2 [5]. **c** Satellite imagery of a segment of Kangaroo Island, South Australia, where drilling operations encountered high hydrogen concentrations reaching 84% [38]

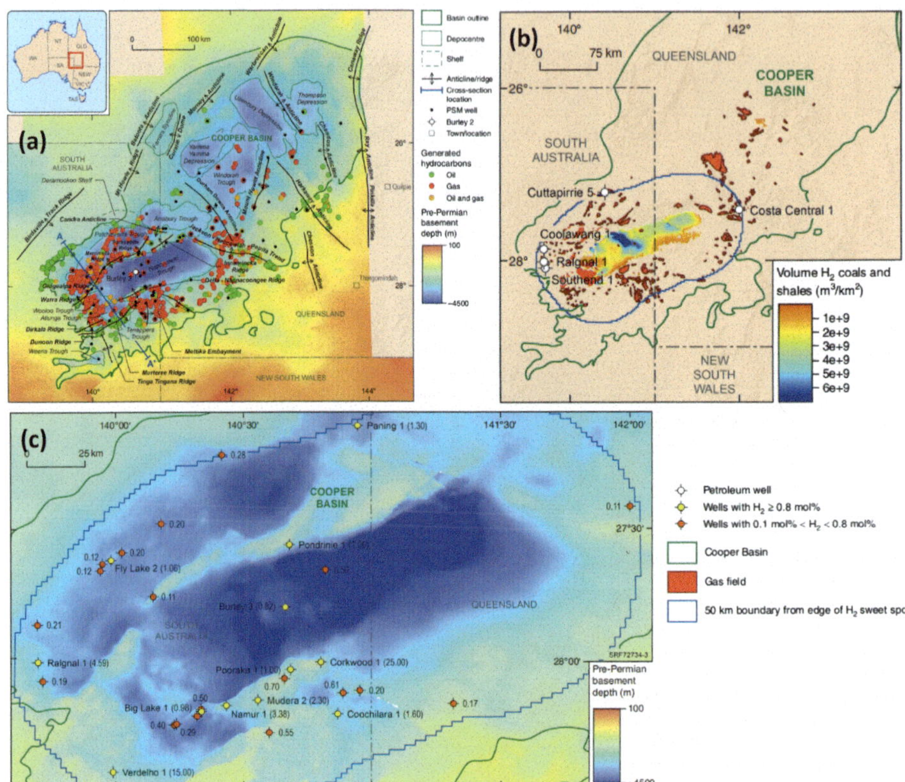

Fig. 3.35 The Hydrogen System of the Cooper Basin. **a** Spatial extent of the Cooper Basin illustrating major depocenters, structural troughs and ridges, along with the distribution of hydrocarbon occurrences. **b** Spatial variation of simulated cumulative hydrogen generation (m^3/km^2) derived from a mid-Patchawarra Formation source rock for free H_2. **c** Locations of wells with reported maximum H_2 concentrations (mol%) where H_2 exceeds 0.1 mol% (reproduced with permission from [52])

within Neoproterozoic sedimentary formations. H_2 accumulations have been documented at depths ranging from shallow intervals (~ 100 m) to substantial subsurface levels ($\leq$ 1800 m). Genetic analyses suggest a deep basement-derived H_2 source, coexisting with elevated concentrations of radiogenic helium, argon, and mantle-associated nitrogen. Doleritic sill complexes function as impermeable barriers to vertical H_2 migration, while aqueous-phase saturation improves reservoir sealing through hydrogen's limited solubility in formation waters [55]. Contemporary gas generation coupled with overpressured reservoirs facilitates the occurrence of H_2-saturated artesian discharges observed in local groundwater extraction wells. Detectable carbon monoxide concentrations imply active H_2-producing reactions, highlighting the system's potential for continuous H_2 regeneration. Economic assessments indicate extraction costs for native H_2 in this region are

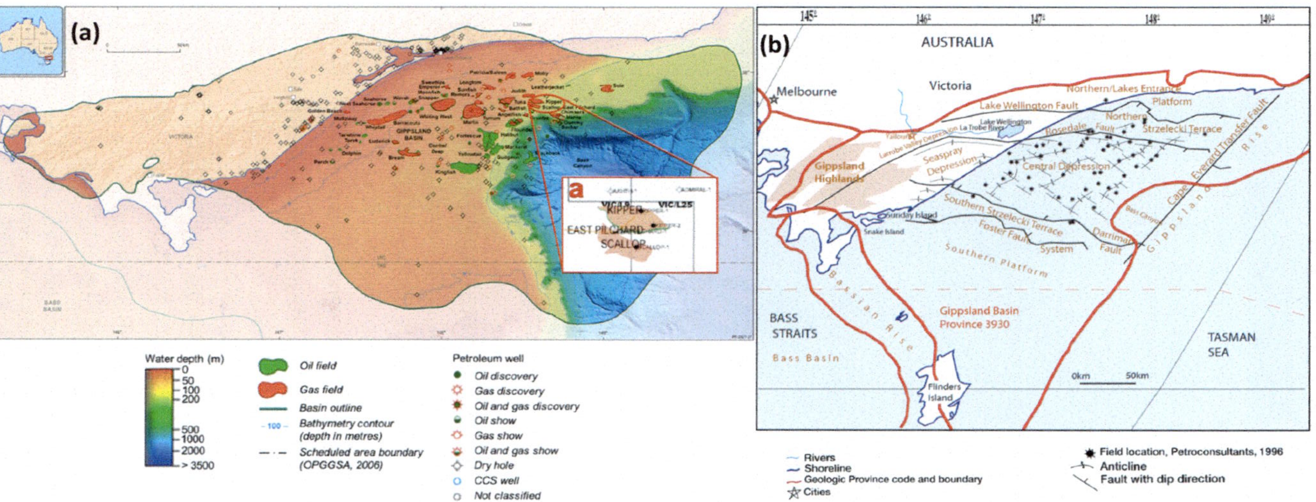

Fig. 3.36 Geological Setting and Petroleum System Elements of the Gippsland Basin. **a** Map of the Gippsland Basin depicting the distribution of oil and gas fields, seabed bathymetry, and petroleum well locations. **b** Map of the Gippsland Basin illustrating the generalized structural framework and major tectonic elements (reproduced with permission from [53])

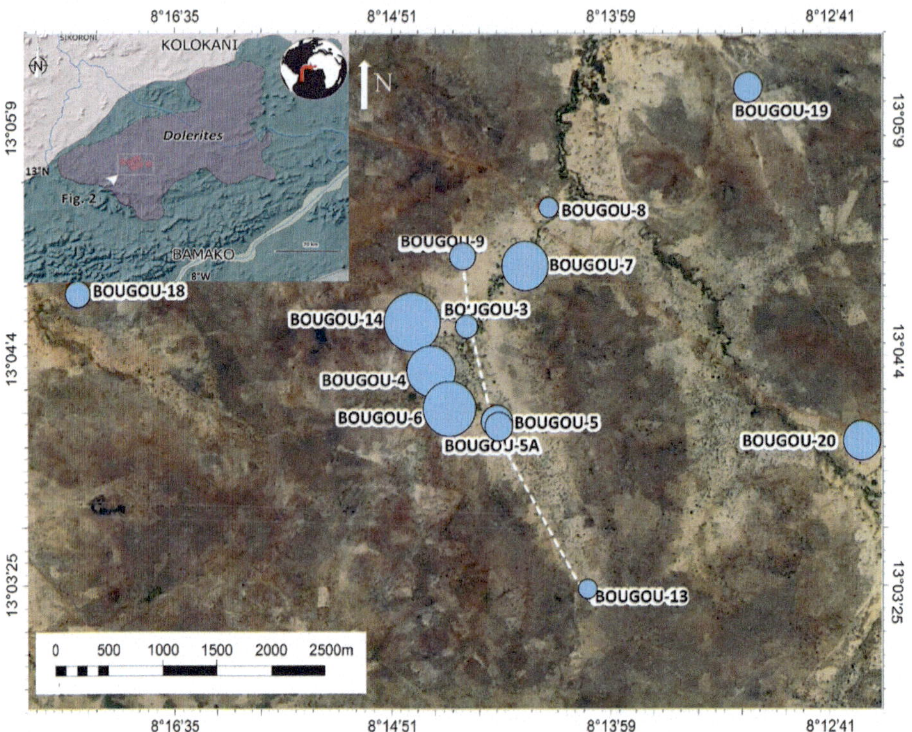

Fig. 3.37 Bourakebougou Field with respective hydrogen accumulations in blue (reproduced with permission from [57])

substantially reduced (2–tenfold) compared to conventional production methods, establishing Bourakebougou as a strategic candidate for sustainable H_2 resource development [56].

3.4.1 Formation Mechanism and Occurrence Conditions

The Bourakebougou field is situated in the southern region of the Taoudeni Mega-Basin in Mali, which comprises a stratigraphic succession ranging from the Archean to the Quaternary. The oldest lithological units, representing the crystalline basement, are exposed along the Leo–Man Ridge located to the south of the study area. These units are predominantly composed of plutonic, volcano-sedimentary, and sedimentary rocks, intruded by magmatic sequences, all of which were structurally influenced by the Eburnean orogeny during the Paleoproterozoic era (approximately 2.2–2.0 Ga) [54].

The results of geochemical analysis indicate that H_2 in the Bourakebougou area of Mali is of inorganic origin, and that H_2 production is related to ancient iron-rich bedrocks and is characterized by sustainable generation. The process by which iron-rich minerals such as olivine come into contact with water and undergo serpentinization is the main source of H_2; Carbonate rocks (mainly dolomite) and sandstone on the basement of the fault block are the main reservoirs of natural hydrogen [54]. Natural hydrogen is present in the reservoir as a free gas above a depth of 800 m and mainly in a water-soluble form below 800 m, with the uppermost carbonate reservoir being the most dominant natural hydrogen producing in the region (Fig. 3.38). The caprock is mainly a mat-like multi-layer intrusive diabase interbed [1].

3.4.2 Reservoir and Preservation

Figure 3.39a presents the geological cross-section of the Bourakebougou gas field [40]. According to the cross-sectional profile, at least five distinct hydrogen-enrich reservoirs are revealed in the Bourakebougou area. This aligns with the findings in the Bougou-6 Well and Bougou-19 Well (Fig. 3.39c-d) [54]. The hydrogen accumulation predominantly occurs in two types of formations: carbonate layers and sandstone layers. Large-scale hydrogen accumulation was found in the dolomitized carbonate rocks of the region, and hydrogen mainly exists in a free state. In contrast, deep hydrogen accumulations are primarily hosted within specific sandstone reservoirs and fractured crystalline basement formations, where hydrogen exists mainly in a dissolved form [55].

The basalt bedrock is the main caprock. The correlation between the thickness of basalt and hydrogen trapping is that the greater the thickness of the basalt, the higher the concentration of hydrogen detected beneath it [57]. The fracture porosity of the basalt bedrock is very low, giving it strong sealing ability to prevent hydrogen from leaking to the surface. Compared to the upper basalt caprock, the fracture porosity of the lower basalt layer is slightly higher, which provides a conduit for hydrogen ascent from deep-seated sources to shallower reservoir formations [55].

3.4.3 Analysis of Production Trends

Beginning in the 2010s, exploratory initiatives and pilot extraction projects have been undertaken to assess the viability of H_2 as an alternative clean energy resource. Subsequent datasets offer a comprehensive overview of evolving production patterns, demonstrating consistent growth in both daily production rates and cumulative output volumes over time (Fig. 3.40) [57].

Analysis criteria of hydrogen production trends includes: (1) Between 2011 and 2020, daily hydrogen (H_2) production increased substantially from 500 Nm^3/day to 2,500 Nm^3/day, representing a fivefold rise due to advances in extraction technologies [54]. (2) Cumulative Production Growth: By 2020, the cumulative H_2 output reached approximately 4.6

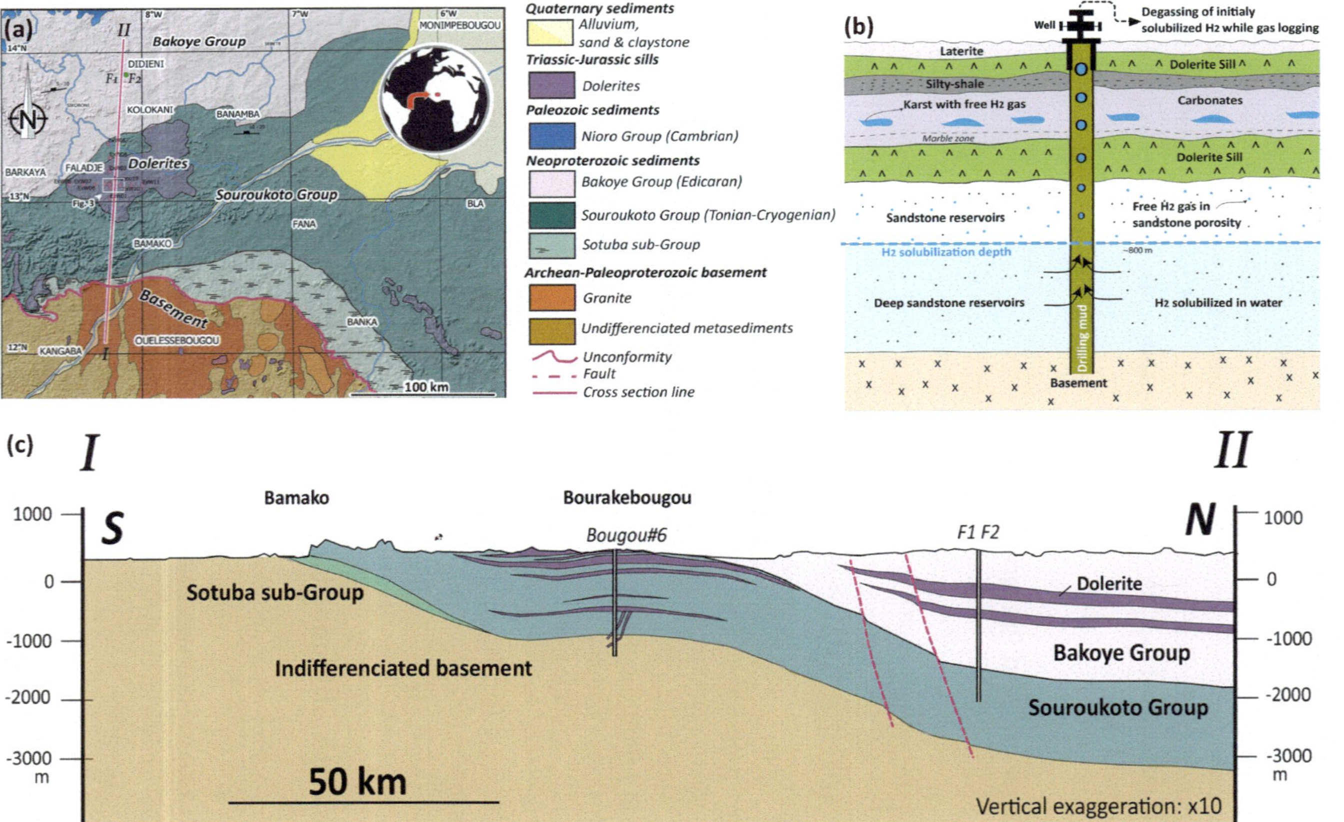

Fig. 3.38 Maps of Bourakebougou well area of Mali. **a** Geological characteristics. **b** stratigraphic profile. **c** Schematic diagram of natural hydrogen occurrence (reproduced with permission from [54])

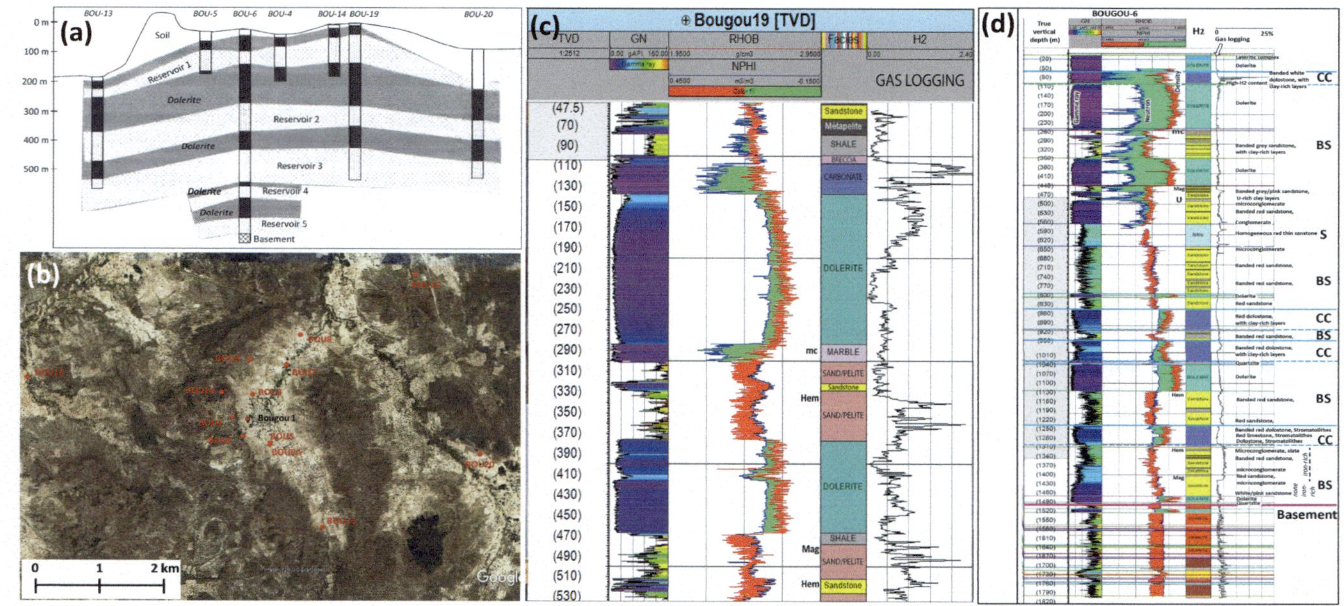

Fig. 3.39 Geological context and multi-method characterization of natural hydrogen occurrences in the Bourakebougou field, Mali. **a** Schematic cross-section of the Bourakebougou gas field in Mali, showing interpreted hydrogen "reservoirs" [40]. **b** Satellite imagery from GoogleEarth of the Bourakebougou region, with well locations and numbers indicated by circles [56]. **c** Datasets used for identifying and characterizing formations and hydrogen accumulation zones between 0 and 530 m depth in Bougou 19. **d** Composite stratigraphic column and electrical log data from Bougou 6 [54]

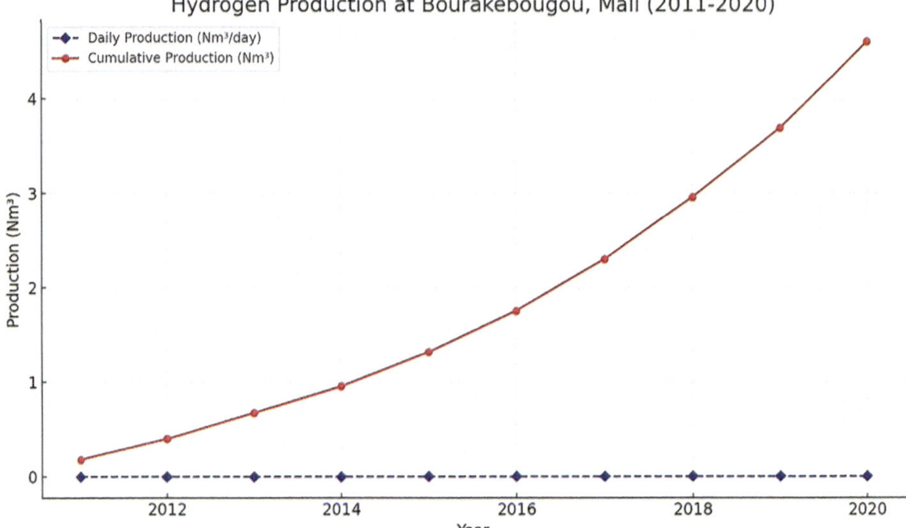

Fig. 3.40 Production rate from 2011 to 2020 in Bourakebougou, Mali (reproduced with permission from [52])

million Nm^3, demonstrating a consistent growth trajectory that highlights the region's viability as a sustainable H_2 reservoir. (3) Purity of Hydrogen: The extracted H_2 from Bourakebougou exhibits exceptionally high purity levels (96–98%), compatible for direct integration into local energy grids with minimal purification requirements [54] (Table 3.6).

3.5 North America

3.5.1 The U.S.A

Carolina bays are geomorphic, consistently oriented, elliptical depressions extensively distributed across the southeastern Atlantic Coastal Plain of eastern North America. While some bays possess continuous raised rims, these typically do not form a complete perimeter and more commonly exhibit a crescentic morphology (Fig. 3.41a–d). Bays of varying sizes commonly exhibit overlapping relationships, with smaller bays often nested within larger ones (Fig. 3.41c). The study focuses on three well-defined Carolina bays, Arthur Road Bay (Fig. 3.41c), Smith Bay (Fig. 3.41b, d), and Jones Lake Bay (Fig. 3.41b), along with four smaller-scale detailed study sites associated with them [58].

At multiple sites including Arthur Road Bay, Smith Bay, Jones Lake Bay, and associated sandpits, elevated hydrogen concentrations were detected, often localized within depressions or along their sandy rims (Fig. 3.41b–d). Maximum H_2 levels exceeded 0.11%

Table 3.6 Hydrogen production at Bourakebougou (2011–2020) [52]

Year	Hydrogen production (Nm^3/day)	Cumulative production (Nm^3)	Hydrogen purity (%)	Utilization
2011	500	182,500	96–98	Local power generation
2012	600	401,500	96–98	Local power generation
2013	750	675,000	96–98	Local power generation
2014	800	957,000	96–98	Local power generation
2015	1000	1,322,500	96–98	Local power generation
2016	1200	1,758,500	96–98	Local power generation
2017	1500	2,304,500	96–98	Local power generation
2018	1800	2,961,500	96–98	Local power generation
2019	2000	3,691,500	96–98	Local power generation
2020	2500	4,606,500	96–98	Local power generation

at Arthur Road locations, while Smith Bay and its intersection zone with an unnamed bay showed peaks > 1200 ppm and > 0.12%, respectively, with concentrations increasing with depth at the peat-to-sand transition (Fig. 3.41d, f). Jones Lake Bay exhibited confined H_2 anomalies up to 815 ppm inside the bay, and a nearby sandpit recorded a high of 0.37% H_2 in clayey sand (Fig. 3.41b, e). Overall, elevated hydrogen was consistently associated with geomorphic structures and sediment type, particularly at rim boundaries or in coarse-grained units [58].

The estimated hydrogen flux rates across the studied areas are as follows: at Smith Bay, daily H_2 flows range from 750 to 1000 m^3/day across an area of 1.14 km^2, equivalent to 660–880 m^3/day/km^2. Arthur Road Bay exhibits a flow rate of 1000–1370 m^3/day over 0.48 km^2, yielding 2240–3060 m^3/day/km^2. In Jones Lake Bay, the estimated emission is 1120–2740 m^3/day over 6.25 km^2, corresponding to 180–440 m^3/day/km^2. Additionally, a minor structure within Jones Lake Bay shows a daily H_2 flow of 21–31 m^3/day over 0.007 km^2, resulting in a significantly higher areal flux of 3000–4400 m^3/day/km^2 [58].

Hydrogen has been detected in Kansas, since the early 1980s, in wells located proximal to the Mid-Continent Rift System. The investigated wells, Heins#1, Scott#1, and Sue Duroche#2, are situated in Geary, Morris, and Riley counties, Kansas, USA,

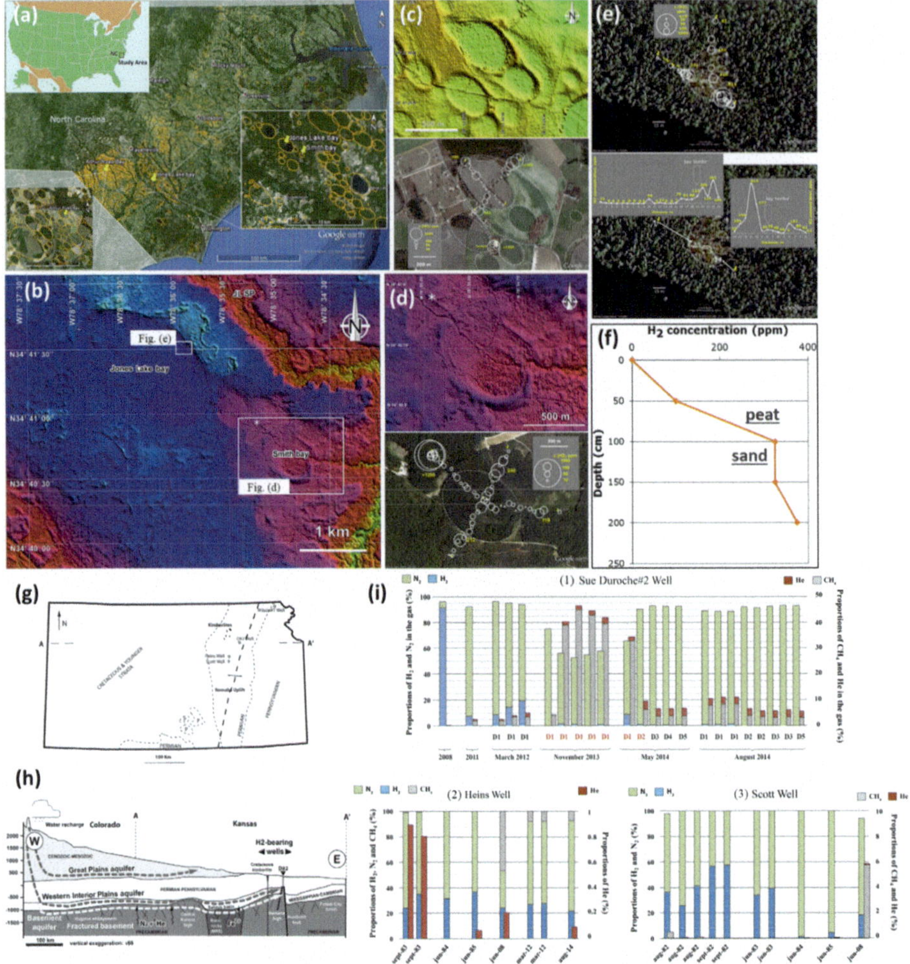

Fig. 3.41 **a** Location of Carolina bays. **b** LiDAR image indicating site locations, including JLSP (Jones Lake Sandpit). The asterisk marks the confluence of Smith Bay and an unnamed bay. **c** Measurements of subsoil H_2 concentration at Arthur Road Bay and Arthur Road Sandpit (lower panel). Dashed lines delineate bay boundaries. The upper panel presents a LiDAR image illustrating the topography corresponding to the area in the lower image. Profile lines were aligned along ditches (visible as dark lines in the photograph). **d** Example of subsoil H_2 concentration measurements from Smith Bay (lower panel). The upper panel shows the corresponding LiDAR image. The asterisk indicates the intersection between Smith Bay and the unnamed bay, with dashed lines outlining Smith Bay. **e** Subsoil H_2 concentrations recorded at Jones Lake Bay, the location of which is provided in Fig. 3.41b. The dashed line denotes the extent of the newly identified structure. **f** Increase in hydrogen concentrations with depth observed at the intersection of Smith Bay and the unnamed bay, the location of which is indicated in Fig. 3.41d (reproduced with permission from [58]). **g–h** Geological and structural schematic map of Kansas illustrating the locations of the Heins#1, Scott#1, and Sue Duroche#2 (D#2) wells. (i) Mole percentage variations of major gas components over time for the (1) Sue Duroche#2, (2) Scott#1, and (3) Heins#1 wells. Sampling dates for the Sue Duroche#2 well are indicated in red for pre-production samples and in black for samples taken during production (reproduced with permission from [59])

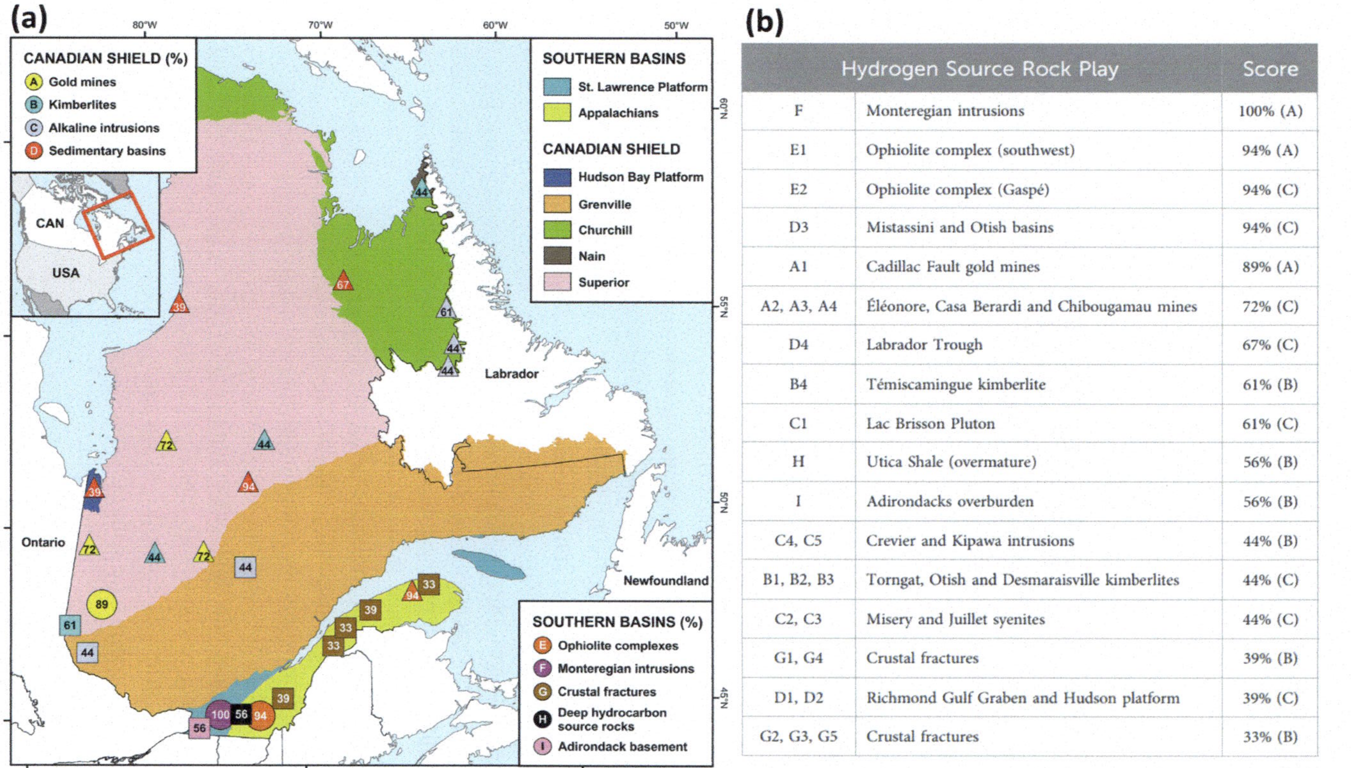

Hydrogen Source Rock Play		Score
F	Monteregian intrusions	100% (A)
E1	Ophiolite complex (southwest)	94% (A)
E2	Ophiolite complex (Gaspé)	94% (C)
D3	Mistassini and Otish basins	94% (C)
A1	Cadillac Fault gold mines	89% (A)
A2, A3, A4	Éléonore, Casa Berardi and Chibougamau mines	72% (C)
D4	Labrador Trough	67% (C)
B4	Témiscamingue kimberlite	61% (B)
C1	Lac Brisson Pluton	61% (C)
H	Utica Shale (overmature)	56% (B)
I	Adirondacks overburden	56% (B)
C4, C5	Crevier and Kipawa intrusions	44% (B)
B1, B2, B3	Torngat, Otish and Desmaraisville kimberlites	44% (C)
C2, C3	Misery and Juillet syenites	44% (C)
G1, G4	Crustal fractures	39% (B)
D1, D2	Richmond Gulf Graben and Hudson platform	39% (C)
G2, G3, G5	Crustal fractures	33% (B)

Fig. 3.42 Distribution and relative prospectivity of potential natural hydrogen source settings within the Quebec sector of the Canadian Shield (reproduced with permission from [60]; Notht that base map was from SIGEOM, 2023). **a** Comparative assessment of candidate hydrogen source-rock domains across Quebec, expressed as percentage-based prospectivity scores assigned to each evaluated area. **b** Integrated ranking results highlighting the relative potential of selected geological domains for natural hydrogen generation within the Canadian Shield and adjacent southern Quebec regions

respectively.These wells align along the Nemaha uplift, several kilometers west of the Humboldt Fault (Fig. 3.42a–b), which truncates both Precambrian basement rocks and lower Paleozoic strata. Kimberlite pipe occurrences are documented in Riley and Marshall counties,approximately 40 km north of the well sites (Fig. 3.42a–b). These ultramafic bodies are often intensely serpentinized and characterized by abundant lizardite and magnetite [59].

In Kansas, multiple wells drilled since 1980, including the previously mentioned Heins#1 and Scott#1, have encountered gas shows rich in H_2 and N_2, with subordinate hydrocarbon components. Temporal variations in the proportions of N_2, H_2, CH_4, and He from the Sue Duroche#2, Scott#1, and Heins#1 wells are illustrated in Fig. 3.42c. These data suggest two distinct sources for hydrogen: a deep-seated origin within the crystalline Precambrian basement, and a shallow, surficial source where H_2 is generated within the well tubing [59].

3.5.2 Canada (Quebec)

In the Quebec portion of the Canadian Shield, a set of seventeen target zones has been recognized and organized into four broad categories according to their geological affinities. Rather than constituting an exhaustive inventory, these zones were selected to capture the range of lithologies and structural settings capable of hosting hydrogen-generating processes. Consequently, the occurrence of natural hydrogen within the Canadian Shield is likely more widespread than indicated by these selected examples.

A wide spectrum of geological settings conducive to the generation of natural hydrogen has been recognized across Quebec, spanning the crystalline terrains of the Canadian Shield to the sedimentary basin systems of the province's southern regions. From this regional survey, twenty-seven prospective domains were selected for further analysis. Each of these identified plays was subsequently examined through a comparative framework based on three key geological parameters. The spatial distribution of the resulting prospectivity rankings is illustrated in Fig. 3.42a, while a synthesized overview of the evaluation outcomes is provided in Fig. 3.42b. Hydrogeochemical investigations conducted in southern Quebec point to localized hydrogen enrichments linked to major tectonic structures. Notable associations have been documented along the Rivière Jacques-Cartier Fault, within the Tracy BrookDelson fault corridor, and in the vicinity of the Aston Fault and the Logan Line. Structural domains characterized by extensional fault architectures, particularly in the first two areas, are regarded as especially promising, as such configurations facilitate vertical permeability and promote the upward transfer of fluids originating at depth [60].

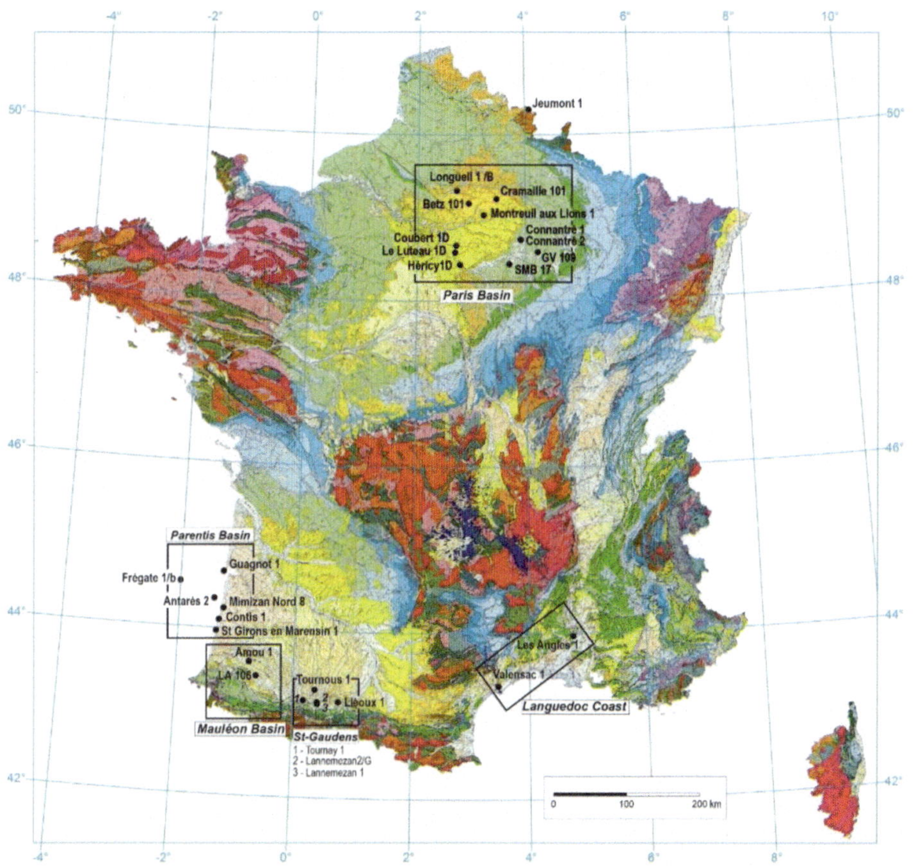

Fig. 3.43 Geological map of France illustrating the distribution of wells in which hydrogen occurrences have been identified through OCR-based detection methods (reproduced with permission from [61])

3.6 Europe

3.6.1 France

France is regarded as one of the countries with the largest currently identified natural hydrogen resources. As an early pioneer in natural hydrogen exploration and research, occurrences of natural hydrogen have been reported from the Pyrenean foreland, the Lorraine region, and the Paris Basin (Fig. 3.43).

3.6.1.1 Pyrenees and Lorraine

One of the area of large natural hydrogen occurrence is situated in southern Gironde, France, adjacent to the Villagrains-Landiras anticline. Small, circular depressions, commonly water-filled and referred to locally as 'lagunes' or 'lagües' (Fig. 3.44a), dominate the landscape. The regional geological framework presents potential for deep H_2 generation, attributed to its proximity to the deeply rifted Parentis Basin and favorable geological conditions linked to Pyrenean orogenic activity. This E-W trending anticlinal structure may serve as a conduit for subsurface fluid migration. The anticline's potential fault network, disrupting Triassic evaporite sequences, may facilitate upward migration of deep-seated fluids through the sedimentary column [59, 62].

Initial gas composition assessments detected a significant anomaly in H_2 levels. Soil gas measurements obtained using the GA-5000 instrument recorded H_2 concentrations as high as 9000 ppm. No discernible spatial correlation between H_2 concentrations and proximity to the lagoons was established. Elevated H_2 levels (approaching 9000 ppm) were detected near circular geological structures; however, these measurements were not uniformly distributed around the lagoon perimeter (Fig. 3.44b). To validate anomalously high H_2 readings, immediate replication of measurements in adjacent areas was performed [62].

According to a representative continuous monitoring profile of H_2 concentrations during a single drilling operation. The trend reveals a rapid surge to saturation levels immediately following the drilling activity, succeeded by a gradual decline over the remaining measurement period, which spans approximately 23 min. This behavior can be attributed to an abrupt emission of H_2, coupled with a lack of significant recharge to sustain elevated concentrations [62].

Two flux chambers, each with an internal volume of 0.0557 m^3, were installed and buried to a depth of 1 m. Gas chromatography (GC) analysis of extracted chamber gases revealed that the apparent 3 ppm H_2 concentration detected in monitoring datasets was artificially generated by temperature-induced drifts in the electrochemical sensors, with diurnal and seasonal fluctuations in sensor readings exhibiting strong correspondence with external air temperature variations (Fig. 3.44c). Continuous monitoring over 12 months at this location failed to detect any measurable H_2 emissions, contrary to elevated H_2 levels documented in adjacent areas prior to, throughout, and following the experimental timeframe [62]. Lefeuvre carried out work on natural H_2 in piedmont of the French Pyrenees and concluded that the deep circulation of water contributes to the generation and transport of natural hydrogen (Fig. 3.44d) [63, 64].

3.6.1.2 Paris Basin

Application of optical character recognition (OCR) technology enabled a systematic analysis of a large number of historical End of Drilling Reports (EDR) from the French CVAGeoDB database. This led to the identification, for the first time, of several drill holes with natural hydrogen (H_2) shows in the Paris Basin [61]. Such shows were often overlooked or misreported (e.g., confused with H_2S) during conventional hydrocarbon exploration in the past. Through manual verification, 11 hydrogen-bearing drill holes

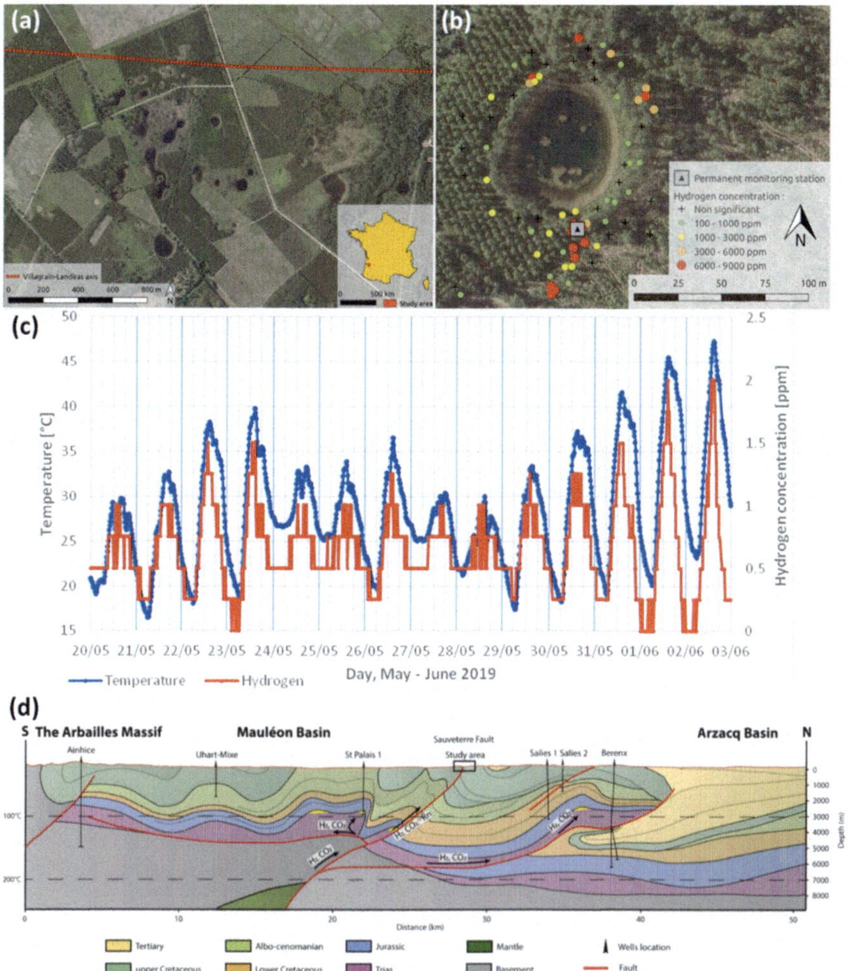

Fig. 3.44 Multiscale Evidence for Natural Hydrogen Seepage in the Villagrains-Landiras Anticline, France. **a** Numerous circular topographic depressions, largely water-saturated, are identifiable in aerial imagery of the area. These features, known as 'lagües' within the study zone, are situated north of the village of Saint-Magne and south of the Villagrains-Landiras structural axis. **b** Soil-gas hydrogen concentrations and sampling locations near the Eaux Belles lagune, situated 2 km north of Saint-Magne. Continuous monitoring was conducted for over one year across the southern segment of the structure. **c** Characteristic oscillatory signal in hydrogen concentration measured by a fixed monitoring station (reproduced with permission from [62]). **d** Schematic diagram of natural hydrogen accumulation in Pyrenean piedmont, France. Natural hydrogen transport via thrust faults in foothill basins: A case study from the North Pyrenean Frontal Thrust (reproduced with permission from [63])

were confirmed in the basin (Fig. 3.45a). Among these, downhole sampling in the Dogger aquifer of the Montreuil Aux Lions 1 well directly measured a hydrogen concentration as high as 52 vol%, representing the highest value recorded in the basin to date [65].

The spatial distribution of hydrogen shows in the Paris Basin exhibits distinct patterns. Discovered hydrogen-bearing wells are predominantly confined to an area of approximately 8,600 km^2 in the central-eastern part of the basin (Fig. 3.45b). Most of these wells align along the NW-SE trending Bray Fault and its associated thrust belt. This spatial arrangement is closely correlated with the distribution of the polarization-corrected vertical gradient of magnetic anomalies extending to depths of 600 m (Fig. 3.45c) [66]. Hydrogen primarily occurs within four specific geological reservoirs: the Lusitanian aquifer, the Dogger aquifer, the Triassic aquifer, and the basement (Fig. 3.45d). This suggests that deep-seated faults such as the Bray Fault likely serve as primary conduits for the upward migration of hydrogen from depth, governing both its transport and local accumulation [61].

The Paris Basin is an intracratonic basin underlain by a Variscan suture zone, which may contain ultramafic and granitic rocks [67]. The detection of excess ^3He in the basin indicates contributions from mantle-derived fluids. Consequently, the genesis of hydrogen is inferred to potentially involve serpentinization and water radiolysis [68, 69]. Overall, the Paris Basin may host a rudimentary yet complete hydrogen system, comprising possible source rocks in the deep subsurface (serpentinized peridotite or radiolytic sources), migration pathways such as the Bray Fault, and potential seals provided by evaporites and claystones within the Triassic and Jurassic sequences [65].

The natural hydrogen shows in the Paris Basin were identified from extensive historical datasets that had been underutilized, revealing a previously overlooked prospective area for natural hydrogen accumulation. This case demonstrates that economically viable natural hydrogen systems may exist in conventional hydrocarbon provinces even in the absence of clear indicators of hydrogen source rocks, such as those typically sourcing from organic-rich strata [65].

3.6.2 Spain: Aragonese Pyrenees

The location of the Aragonese Pyrenees is shown in Fig. 3.46a, and its formation is due to the Iberian rotation and crustal extension of the Middle Cretaceous Pyrenees (Fig. 3.46b). The geology is shown in Fig. 3.46d, e, with the H_2 anomaly detected in the Northern Pyrenees (Fig. 3.46c–g), and Lefeuvre et al. [64] documented natural hydrogen seepage along the North Pyrenean Fault zone on the southern margin of the Aquitaine Basin (marked in red) (Fig. 3.46c). Note the symmetrical position of the Permit area (blue) along the northern margin of the Ebro Basin within the South Pyrenean zone. The occurrence of a thick Mesozoic to Tertiary sedimentary cover in the South Pyrenees enhances hydrogen entrapment, in contrast to the North Pyrenees, where this cover is largely absent (Fig. 3.46e) [70].

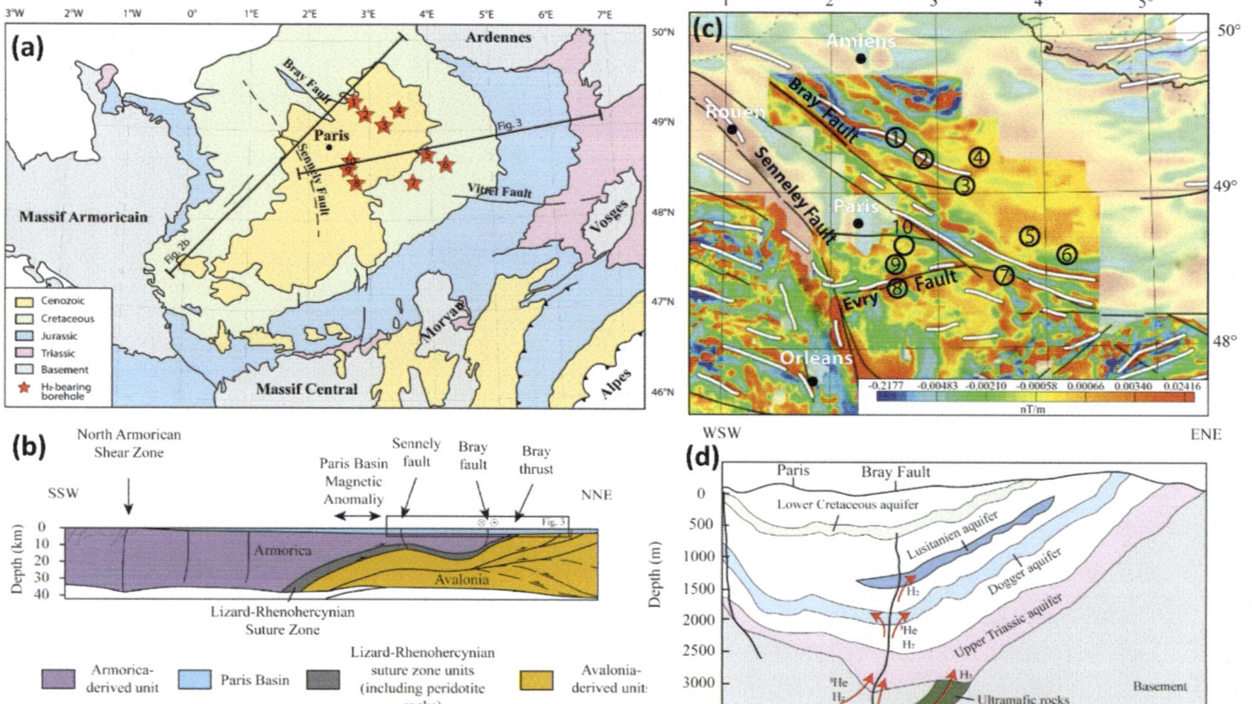

Fig. 3.45 The natural hydrogen system in the Paris Basin. **a** The structural map of the Paris Basin displays the main units and the surrounding crystalline massifs. In this map, the red stars represent the wells showing evidence of the presence of H_2, as identified using the Optical Character Recognition algorithm developed in the study 1—Longeuil; 2—Betz; 3—Montreuil Aux Lions; 4—Cramaille; 5—Connantre 1 and 2; 6—Grandville; 7—Saint Martin de Bossenay; 8—Hericy; 9—Le Luteau; 10—Coubert. **b** Crustal-scale cross-section through the Variscan orogenic system and the Paris Basin. **c** Map of the vertical gradient of the magnetic anomaly reduced to the pole and extended to 600 m up. Black lines correspond to major discontinuities, white lines to magnetic axes and the black circle corresponds to the H_2-bearing well 1—Longeuil; 2—Betz; 3—Montreuil Aux Lions; 4—Cramaille; 5—Connantre 1 and 2; 6—Grandville; 7—Saint Martin de Bossenay; 8—Hericy; 9—Le Luteau; 10—Coubert. **d** A schematic cross-section of the Paris Basin, with the positions of the four main aquifers where H2 was detected and the putative H_2 system beneath the Paris Basin (modified from [61, 65])

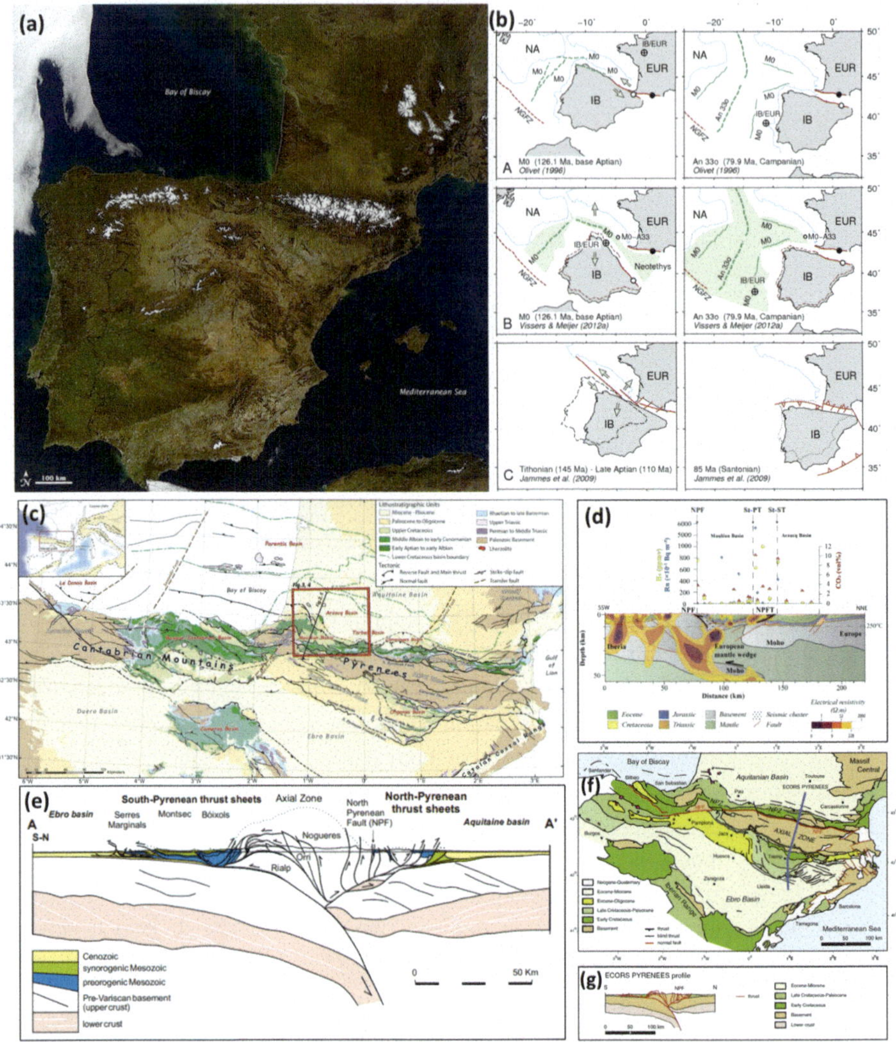

Fig. 3.46 The Mid-Cretaceous Pyrenees: Geographical Setting and Geodynamic Model. **a** Geographical location of the Aragonese Pyrenees. **b** The Iberian rotation and crustal extension of the Middle Cretaceous Pyrenees show intent [70]. The Natural Hydrogen System of the Pyrenees: Seepages, Anomalies, and Trapping Mechanisms: **c** Natural hydrogen (H_2) seepage occurrences are documented along the Pyrenean Fault System at the southern periphery of the Aquitaine Basin. **d** H_2 anomalies detected in the Northern Pyrenees. **e** Mechanism of H_2 trapping by Mesozoic/Tertiary thick caprock sediments in the Southern Pyrenees. **f–g** Geology profile of the Pyrenees (reproduced with permission from [70])

3.7 Brazil

The São Francisco Basin is a north–south (N–S) trending double foreland basin developed on the São Francisco Craton. It is flanked by the Brasília and Araçuaí orogenic belts and subdivided by structural highs and the Pirapora Aulacogen (Fig. 3.47a). Looking at the location of the kitch ens and the H2 shows and indicators, Some scholars proposed several possible migration routes in the basin, from long and short distances, both vertical and horizontal (Fig. 3.48) [71]. Hydrogen measurements using Parhys H_2 detectors revealed subsurface gas extraction from -80 cm depth, yielding characteristic bell-shaped concentration profiles, though instrumental limits may truncate peak values (Fig. 3.47b). By utilizing an array of detectors, the spatiotemporal analysis of H_2 emissions revealed that primary sources are predominantly located along the vegetated periphery of geomorphological structures, Intermittent high-magnitude emissions were also detected near the central region (Fig. 3.47c). Migration pathways from leakage sources to the surface are influenced by soil heterogeneity, where prolonged migration results in delayed peak H_2 accumulation at 80 cm depth and reduced concentrations due to dispersion and microbial consumption [72].

3.8 Oman

The study area encompasses the Western Hajar Mountains in northern Oman. Figure 3.49a illustrates the regional geological framework, including the spatial distribution of measurement sites. The ophiolitic sequences and their underlying basement rocks in this region underwent extensive secondary deformation, resulting in the development of a prominent basement antiform within the Jebel Akhdar dome. A well-exposed outcrop of Precambrian sedimentary basement rocks is documented in this area (Fig. 3.49b). Figure 3.49b also presents a synthesized structural cross-section of the investigated region, highlighting its geological architecture. Boreholes were drilled through massive peridotite formations crosscut by fractures and diabase dykes (Fig. 3.49c-d). Elevated H_2 concentrations were detected in nearly all fracture systems, with measurements (derived from a minimum of 40 sampling points) yielding an average concentration of 73 ppm and localized peaks exceeding 650 ppm (Fig. 3.49b) [73].

Moderate H_2 concentrations ($H_2 < 30$ ppm) were detected in the recent sediments of the Batinah coastal plain, composed of loess and alluvial deposits overlying ophiolitic formations. These H_2 levels exhibited a gradual decline toward the coastline, diminishing to undetectable amounts (Fig. 3.49b). In contrast, the subjacent Hawasina marine sediments consistently displayed higher hydrogen concentrations than the overlying serpentinized peridotites (Fig. 3.49b). Within Mesozoic carbonate rocks and the Proterozoic to Paleozoic crystalline basement, H_2 concentrations averaged 10 ppm across six sampling sites, reaching a peak of 23 ppm (Fig. 3.49b). Notably, the highest H_2 enrichments were observed in

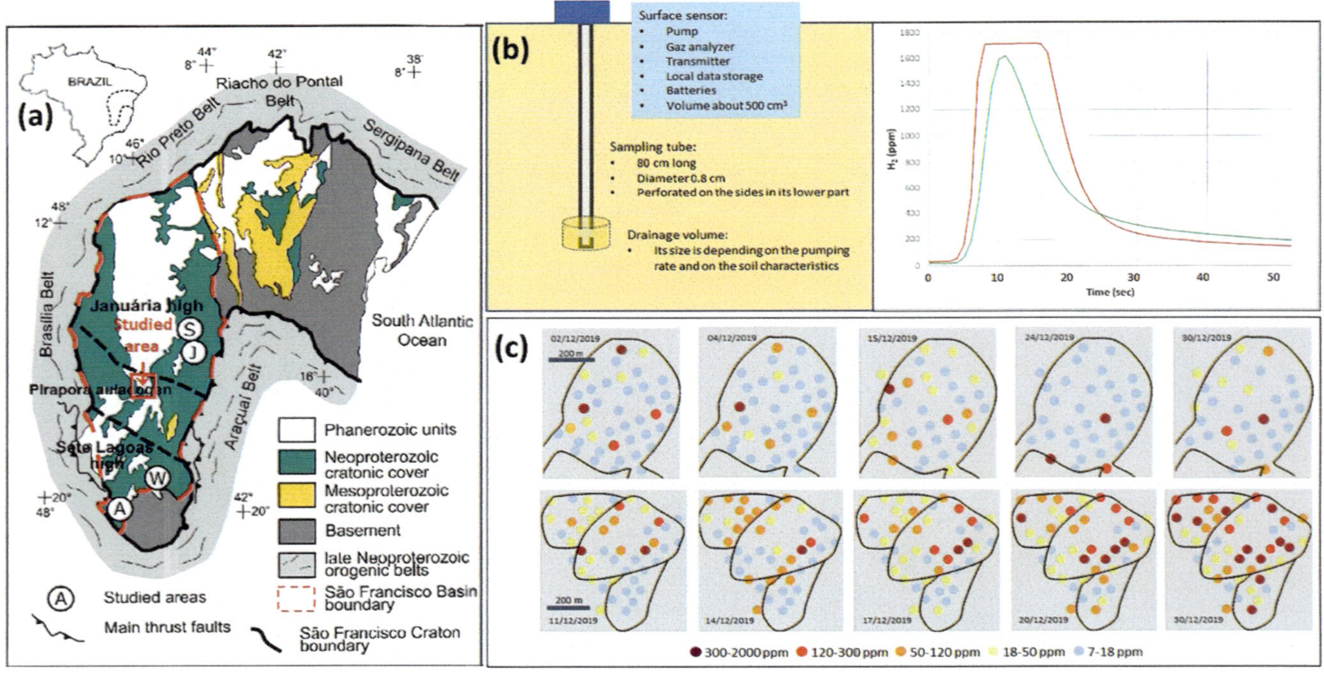

Fig. 3.47 Hydrogen gas emission monitoring in the circular structures of the São Francisco Basin: geological setting, methodology, and time-series examples. **a** Structural map of the São Francisco Basin illustrating the adjacent orogenic belts and the study area location. **b** Left: monitoring setup; right: schematic representation of hydrogen measurements taken during hourly 2-min operational intervals. The figure displays the first 50 s of measurement, beyond which the signal decreases monotonically. Measurements occasionally show unsaturated signals (green curve) or clear saturation (red curve). **c** Diurnal variation of average emissions in Campinas (top) and Baru (bottom) illustrated for five selected days in December 2019 (reproduced with permission from [72])

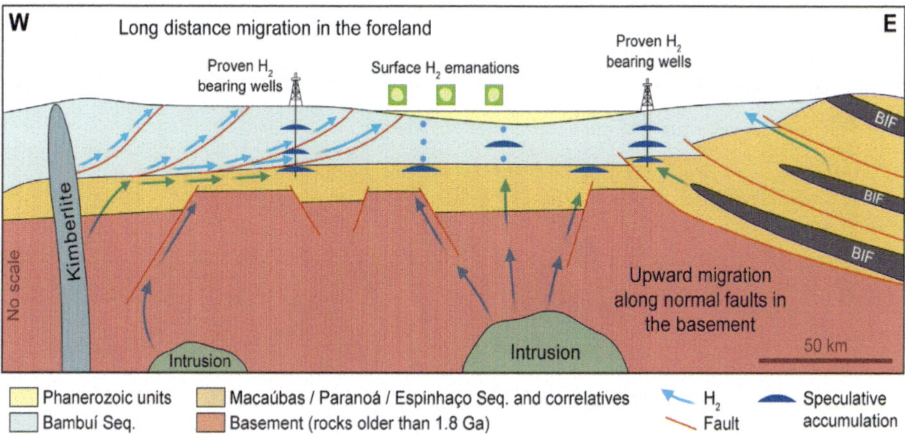

Fig. 3.48 Schematic regional geological profile crossing central-southern São Francisco Basin with the elements of H_2 system. Lithostratigraphic units and faults interpreted based on composite of seismic lines and wells. Basement H_2 generating rocks based on regional geological understanding and magnetic anomalies (reproduced with permission from [71])

Proterozoic to Permian metasedimentary sequences, where measurements from 15 sites yielded an average of 443 ppm, with a remarkable maximum concentration of 3400 ppm (Fig. 3.49b) [73].

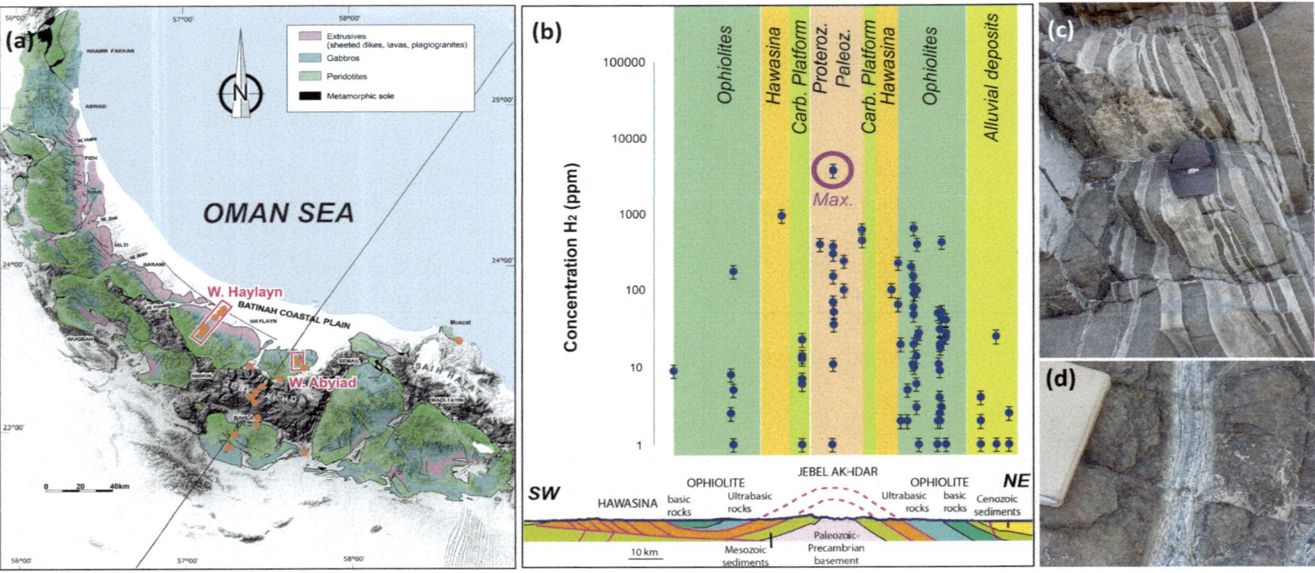

Fig. 3.49 Geological Controls on Natural Hydrogen Surface Expressions in Oman. **a** Geological map of the study area showing the locations of sampling sites (indicated by orange dots). **b** Above: Measured hydrogen concentrations in surface soil and fractured rock samples, displayed in relation to their stratigraphic position within the formations illustrated in the geological map (Fig. 3.49a); Below: Interpreted synthetic cross-section across the study area, aligned with the sampling traverse (black line in Fig. 3.49a). Vertical scale on the cross-section is exaggerated. **c** Representative occurrence of diabase dikes intruding peridotites, where diffuse hydrogen fluxes were recorded. **d** Example of a fracture filled with carbonate cement (dolomite and magnesite), through which diffuse hydrogen emission was detected (reproduced with permission from [73])

References

1. Dou, L., et al. 2024. Global natural hydrogen exploration and development situation and prospects in China (in Chinese). *Lithologic Reservoirs* 36 (02): 1–14.
2. Zhang, Y., et al. 2024. Thoughts and suggestions on the proactive planning for the development and utilization of natural hydrogen (in Chinese). *International Petroleum Economics* 32 (S1): 84–91.
3. Truche, L., T.M. McCollom, and I. Martinez. 2020. Hydrogen and abiotic hydrocarbons: molecules that change the world. *Elements: An International Magazine of Mineralogy, Geochemistry, and Petrology* 16 (1): 13–18.
4. Zgonnik, V. 2020. The occurrence and geoscience of natural hydrogen: A comprehensive review. *Earth-Science Reviews* 203: 103140.
5. Boreham, C.J., et al. 2021. Hydrogen in Australian natural gas: occurrences, sources and resources. *The Australian Energy Producers Journal* 61 (1): 163–191.
6. Fang, D., et al. 2024. Genesis of natural hydrogen and progress in global hydrogen resource exploration (in Chinese). *Journal of Chengdu University of Technology (Science & Technology Edition)* 1–24.
7. Czado, K. 2024. Natural hydrogen: the race to discovery and concept demonstration.
8. Blay-Roger, R., et al. 2024. Natural hydrogen in the energy transition: Fundamentals, promise, and enigmas. *Renewable and Sustainable Energy Reviews* 189: 113888.
9. Huang, J. 1960. A preliminary summary of the basic features of Chinese geological formations (in Chinese). *Acta Geologica Sinica* 1960 (01): 1–31+135.
10. Sun, L., et al. 2024. Geological survey and study of hydrogen-rich natural gas in Songliao Basin (in Chinese). *Petroleum Geology & Oilfield Development in Daqing* 43 (03): 7–16.
11. Jin, Z., et al. 2024. Discovery of anomalous hydrogen leakage sites in the Sanshui Basin. *South China. Science bulletin* 69 (9): 1217–1220.
12. Shuai, Y., et al. 2009. Geochemical evidence for strong ongoing methanogenesis in Sanhu region of Qaidam Basin (in Chinese). *Scientia Sinica (Terrae)* 39 (06): 734–740.
13. Han, S., et al. 2022. Hydrogen-rich gas discovery in continental scientific drilling project of Songliao Basin, Northeast China: New insights into deep Earth exploration (in Chinese). *Science Bulletin* 67 (10): 1003–1006.
14. Lin, L., et al. 2009. Geochemistry and accumulation of the upper palaeozoic natureal gas in the sulige gas field, ordos basin (in Chinese). *Sedimentary Geology and Tethyan Geology* 29 (02): 77–82.
15. Zhang, S., et al. 2007. Exploration of the gas source of a large gas field cluster with high hydrogen sulphide content in the Feixianguan Formation, northeastern Sichuan Basin (in Chinese). *Chinese Science Bulletin* S1: 86–94.
16. Yu, C. 2012. Analysis of shale gas formation conditions and resource potential of the Lower Silurian in Southeast Sichuan area. SouthWest Petroleum University.
17. Hao, Y., et al. 2020. Origin and evolution of hydrogen-rich gas discharges from a hot spring in the eastern coastal area of China. *Chemical Geology* 538: 119477.
18. Jin, Z., et al. 2002. Primary study of geochemical features of deep fluids and their effectiveness on oil/gas reservoir formation in sedimental basins (in Chinese). *Earth science* 27 (6): 659–665.
19. Gao, Q. 2004. Volcanic hydrothermal activities and gas-releasing characteristics of the Tianchi Lake region, Changbai Mountains (in Chinese). *Acta Geoscientica Sinica* 03: 345–350.
20. ShangGuan, Z., C. Bai, and M. Sun. 2000. Characteristics of gas release from modern mantle-sourced magmas in the Tengchong Hot Sea area (in Chinese). *Scientia Sinica (Terrae)* 04: 407–414.

21. Li, X., Y. Liu, and J. Wen. 2002. Geochemical characteristics of the natural gas from Well Wulong-1, Chuxiong Basin, and its geological significance (in Chinese). *Natural Gas Industry* 2002 (05): 16–19+11.
22. Qin, C., et al. 2017. Reservoir characteristics of organic-rich mudstone of the Niutitang Formation in northern Guizhou (in Chinese). *Journal of Southwest Petroleum University (Science & Technology Edition)* 39 (04): 13–24.
23. Kang, J., et al. 2020. Analysis of geochemical characteristics of hydrogen in fault gases in the north section of Yilan-Yitong fault (in Chinese). *Seismological and Geomagnetic Observation and Research* 41 (04): 111–120.
24. Wang, L., et al. 2024. The occurrence pattern of natural hydrogen in the Songliao Basin, P.R. China: Insights on natural hydrogen exploration. *International Journal of Hydrogen Energy* 50: 261–275.
25. Han, S., et al. 2024. Geochemistry and origins of hydrogen-containing natural gases in deep Songliao Basin, China: Insights from continental scientific drilling. *Petroleum Science* 21 (2): 741–751.
26. Meng, Q., et al. 2015. Distribution and geochemical characteristics of hydrogen in natural gas from the Jiyang Depression, Eastern China. *Acta Geologica Sinica (English Edition)* 89 (5): 1616–1624.
27. Jin, Z., et al. 2002. Deep fluid activities and their effects on generationof hydrocarbon in Dongying depression (in Chinese). *Petroleum Exploration and Development* 02: 42–44.
28. Liu, J., et al. 2019. Influences of the deep fluid on organic matter during the hydrocarbongeneration and evolution process (in Chinese). *Natural Gas Geoscience* 30 (4): 478–492.
29. Shen, B., et al. 2009. Noble gas geochemistry of CO_2 gas pool in Gaoqing-Pingnan Fault Zone, Jiyang Depression (in Chinese). *Geological Journal of China Universities* 15 (04): 537–546.
30. Liu, Q., et al. 2017. Effects of deep CO_2 on petroleum and thermal alteration: The case of the Huangqiao oil and gas field. *Chemical Geology* 469: 214–229.
31. Xu, Y., et al. 2013. Study on mantle plume and large igneous provinces in China: An overview and perspectives (in Chinese). *Bulletin of Mineralogy, Petrology and Geochemistry* 32 (01): 25–39.
32. Liu, Q. Y., et al. 2008. Hydrogen isotope composition of natural gases from the Tarim Basin and its indication of depositional environments of the source rocks. *Science in China Series D-Earth Sciences* 51 (2): 300–311.
33. Yang, S., et al. 2007. Permian bimodal dyke of Tarim Basin, NW China: Geochemical characteristics and tectonic implications. *Gondwana Research* 12 (1): 113–120.
34. Yin, F., G. Pan, and Z. Sun. 2021. Genesis and evolution of the structural systems during the cenozoic in the Sanjiang orogenic belt, Southwest China (in Chinese). *Sedimentary Geology and Tethyan Geology* 41 (02): 265–282.
35. Guo, Q., and Y. Wang. 2012. Geochemistry of hot springs in the Tengchong hydrothermal areas, Southwestern China. *Journal of Volcanology and Geothermal Research* 215–216: 61–73.
36. Liu, H., et al. 2018. Crustal footprint of the Hainan plume beneath Southeast China. *Journal of Geophysical Research: Solid Earth* 123 (4): 3065–3079.
37. Yuan, X. 2019. The record of cenozoic magmatism in Sanshui Basin and its relationship with the early tectonic evolution stage of the South China Sea. China University of Geosciences (Beijing) Beijing, China.
38. Reza, R. 2021. Assessment of natural hydrogen systems in Western Australia. *International Journal of Hydrogen Energy* 46 (66): 33068–33077.
39. Haines, P. W., et al. 2009. Isotopic and geochemical characterisation of the Cambrian Kanmantoo Group, South Australia: Implications for stratigraphy and provenance. *Australian Journal of Earth Sciences* 56 (8): 1095–1110.

40. Vitaly, V., and R. Reza. 2022. Natural deep-seated hydrogen resources exploration and development: Structural features, governing factors, and controls. *Journal of Energy and Natural Resources* 11 (3): 60–81.
41. Department of Mines Industry Regulation and Safety, Meda-1 well. Geological Survey of Western Australia.
42. Leila, M., K. Loiseau, and I. Moretti. 2022. Controls on generation and accumulation of blended gases (CH_4/H_2/He) in the Neoproterozoic Amadeus Basin, Australia. *Marine and Petroleum Geology* 140.
43. Emanuelle, F., et al. 2021. Natural hydrogen seeps identified in the North Perth Basin, Western Australia. *International Journal of Hydrogen Energy* 46 (61): 31158–31173.
44. Vidavskiy, V., et al. 2024. Natural hydrogen in the northern Perth Basin, WA Australia: Geospatial analysis and detection in soil gas for early exploration. *Journal of Energy and Natural Resources* 13 (2): 90–113.
45. Esri, M. 2021. E.g., USDA FSA, USGS, aerogrid, IGN, IGP, and the GIS user community, word imagery.
46. Mory, A.J., and R.P. Iasky. 1996. Stratigraphy and structure of the onshore northern Perth Basin, Western Australia. Western Australia. *Geological Survey. Report* 46, 101.
47. Cassidy, K.F., et al. 2006. A revised geological framework for the Yilgarn craton, Western Australia.
48. Léo, A., et al. 2023. Natural hydrogen seeps or salt lakes: how to make a difference? Grass Patch example, Western Australia. *Frontiers in Earth Science* 11.
49. Wingate, M. 2017. Mafic dyke swarms and large igneous provinces in Western Australia get a digital makeover.
50. Davies, K., et al. 2024. A natural hydrogen seep in Western Australia: Observed characteristics and controls. *Science and Technology for Energy Transition* 79.
51. Bourdet, J., et al. 2023. Natural hydrogen in low temperature geofluids in a Precambrian granite, South Australia. Implications for hydrogen generation and movement in the upper crust. *Chemical Geology* 638.
52. Boreham, C., et al. 2023. Modelling of hydrogen gas generation from overmature organic matter in the Cooper Basin, Australia. *The APPEA Journal* 63: S351–S356.
53. Kamalrulzaman, K. N., M. R. Shalaby, and M. A. Islam. 2023. Source rock characteristics and 1D basin modeling of the Lower Cretaceous Latrobe Group, Gippsland Basin, Australia. *Geoenergy Science and Engineering* 225: 211674.
54. Maiga, O., et al. 2023. Characterization of the spontaneously recharging natural hydrogen reservoirs of Bourakebougou in Mali. *Scientific reports* 13 (1): 11876–11876.
55. Su, Y., et al. 2024. Natural hydrogen exploration: A case study of hydrogen wells in the Mali gas field in Africa and global advances (in Chinese). *Oil & Gas Geology* 45 (5): 1502–1510.
56 Prinzhofer, A., C. S. T. Cissé, and A. B. Diallo. 2018. Discovery of a large accumulation of natural hydrogen in Bourakebougou (Mali). *International Journal of Hydrogen Energy* 43 (42): 19315–19326.
57. Maiga, O., et al. 2024. Trapping processes of large volumes of natural hydrogen in the subsurface: The emblematic case of the Bourakebougou H_2 field in Mali. *International Journal of Hydrogen Energy* 50: 640–647.
58. Zgonnik, V., et al. 2015. Evidence for natural molecular hydrogen seepage associated with Carolina bays (surficial, ovoid depressions on the Atlantic Coastal Plain, Province of the USA). *Progress in Earth and Planetary Science* 2(1).
59. Guélard, J., et al. 2017. Natural H_2 in Kansas: Deep or shallow origin? *Geochemistry, Geophysics, Geosystems* 18 (5): 1841–1865.

60. Séjourné, S., et al. 2024. Potential for natural hydrogen in Quebec (Canada): a first review. *Frontiers in Geochemistry* 2: 1351631
61. Lefeuvre, N., et al. 2024. Characterizing Natural Hydrogen Occurrences in the Paris Basin Using OCR-Enhanced Well Database Studies. Authorea Preprints.
62. Paul, H., et al., Hydrogen gas in circular depressions in South Gironde, France: Flux, stock, or artefact? *Applied Geochemistry* 127 (prepublish): 104928.
63. Yin, L., et al. 2024. Origins and accumulation characteristics of large-scale generation of natural hydrogen (in Chinese). *Lithologic Reservoirs* 36 (06): 1–11.
64. Lefeuvre, N., et al. 2022. Natural hydrogen migration along thrust faults in foothill basins: The North Pyrenean Frontal Thrust case study. *Applied Geochemistry* 145: 105396.
65. Lefeuvre, N., et al. 2024. Characterizing natural hydrogen occurrences in the Paris basin from historical drilling records. *Geochemistry, Geophysics, Geosystems* 25 (5): e2024GC011501.
66. Baptiste, J., et al. 2016. Mapping of a buried basement combining aeromagnetic, gravity and petrophysical data: The substratum of southwest Paris Basin, France. *Tectonophysics* 683: 333–348.
67. Averbuch, O., and C. Piromallo. 2012. Is there a remnant Variscan subducted slab in the mantle beneath the Paris basin? Implications for the late Variscan lithospheric delamination process and the Paris basin formation. *Tectonophysics* 558: 70–83.
68. Hirose, T., S. Kawagucci, and K. Suzuki. 2011. Mechanoradical H2 generation during simulated faulting: Implications for an earthquake-driven subsurface biosphere. *Geophysical Research Letters* 38 (17).
69. Geymond, U., et al. 2023. Reassessing the role of magnetite during natural hydrogen generation. Frontiers in Earth Science. Volume 11.
70. Atkinson, C., et al. 2023. Geological setting of natural "gold" hydrogen in the Pyrenees and implications for exploration worldwide.
71. de Freitas, V.A., et al. 2024. Natural hydrogen system evaluation in the São Francisco Basin (Brazil). *Science and Technology for Energy Transition* 79.
72. Isabelle, M., et al. 2020. Long-term monitoring of natural hydrogen superficial emissions in a Brazilian cratonic environment. Sporadic large pulses versus daily periodic emissions. *International Journal of Hydrogen Energy* 46(5).
73. Zgonnik, V., et al. 2019. Diffused flow of molecular hydrogen through the Western Hajar mountains, Northern Oman. *Arabian Journal of Geosciences* 12 (3): 71.

4 Genesis and Hydrogen Source

Natural hydrogen (H_2), widely known as white hydrogen, is generated and accumulated through specific geological processes occurring within distinct tectonic settings [1]. The key mechanisms including serpentinization, where peridotite and other ultramafic rocks react with water at elevated temperatures, releasing H_2 as a byproduct, prevalent in geologically active regions like mid-ocean ridges, ophiolite complexes, and subduction zones, as well as radiolytic decomposition driven by ionizing radiation from uranium, thorium, and other radioactive elements, which dissociates water molecules into H_2 and O_2, particularly within stable cratonic regions such as Precambrian shields where prolonged geological stability facilitates radiolysis over extended timescales [2]. These accumulations predominantly occur in ultramafic lithologies, crystalline basement structures, and organic-rich sedimentary basins, which serve as effective subsurface reservoirs either retaining H_2 within structural traps or permitting gradual migration toward the surface, as evidenced by significant hydrogen seeps documented globally in tectonically active zones including continental rifts, cratonic margins, and volcanic provinces, with notable resources in Australia, China, the North America, Europe and etc. One typical example is the high-purity hydrogen concentrations found in the Bourakebougou field in Mali, which underscore the commercial potential of natural hydrogen extraction [3]. Understanding the geologic controls on hydrogen generation, migration, and entrapment is therefore essential for delineating prospective reservoirs and developing targeted exploration strategies, prompting this review to examine the formation mechanisms and structural contexts of natural hydrogen systems, with particular emphasis on serpentinization-driven environments and radiolytically active cratonic regions, to assess their viability within the future energy landscape.

Y. Yuan et al., *Natural Hydrogen*, Synthesis Lectures on Renewable Energy Technologies, https://doi.org/10.1007/978-3-032-14469-0_4

4.1 Genesis and Hydrogen Source of Natural Hydrogen

Natural hydrogen originates from a range of geological and biological processes occurring over prolonged geological timescales. This section examines the geological mechanisms underlying H_2 formation, migration, and accumulation, along with the entrapment processes that enable its subsurface retention [4–6]. Geological Processes for H_2 Accumulation involves: Genesis and key mechanisms of hydrogen source, including serpentinization, radiolysis, microbial processes, and mantle degassing. The specific geological environment governs which mechanism predominates, with each process generating H_2 in a distinct manner (Fig. 4.1).

The geological environment conducive to the accumulation of natural hydrogen is critical for Natural Hydrogen. Unlike H_2 produced through artificial processes such as electrolysis or methane reforming, natural hydrogen is produced through various geologic processes within the Earth's crust and becomes concentrated in favorable subsurface reservoir formations (Fig. 4.2) [1]. Understanding the geological conditions that foster the formation and trapping of H_2 is essential for identifying economically viable natural hydrogen. This review provides an overview of the geological processes, rock formations, and structural traps that influence the accumulation of natural hydrogen.

For H_2 to accumulate in commercially viable quantities, key favorable geological conditions and structures are required [1, 4, 7–9]: (i) Crystalline Basement Rocks: The crystalline basement is a crucial environment for H_2 generation and accumulation. These ancient, stable rock formations (composed primarily of granites, gneisses, and other

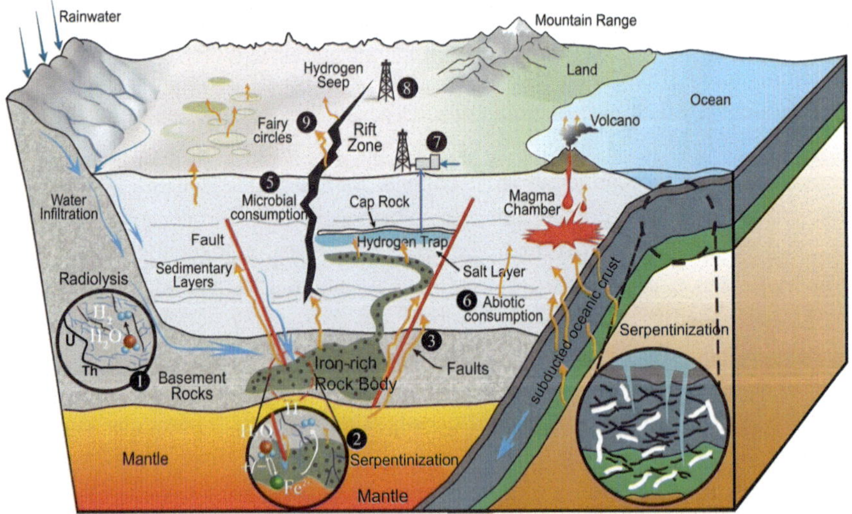

Fig. 4.1 The geological processes involved in the generation of natural hydrogen (modified from [10, 4])

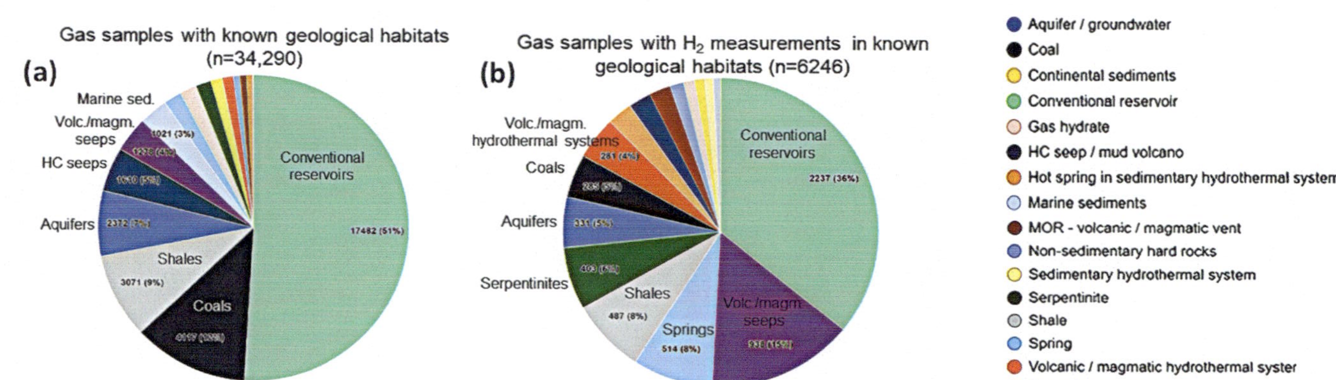

Fig. 4.2 Pie charts illustrating the quantity and distribution of gases from various geological settings. **a** Within the full dataset of samples with documented geological contexts. **b** Within the subset of samples for H_2 concentrations have been analytically determined from [11]

igneous/metamorphic rocks) are often rich in uranium and thorium, making them ideal settings for H_2 production via radiolysis. Additionally, the fractured nature of basement rocks allows for H_2 migration and accumulation in structural traps. (ii) Mid-Ocean Ridges and Ophiolite Complexes: Ultramafic rocks exposed at mid-ocean ridges or preserved as ophiolites are key locations for serpentinization. These areas provide abundant sources of olivine and pyroxene, which react with water to generate H_2. Additionally, faults and fractures within these rocks create pathways for H_2 to migrate toward the surface, where it can be trapped in impermeable layers. (iii) Stable Continental Cratons: Cratonic regions, such as the interior of the African, Brazilian, or Australian shields, are also considered favorable environments for natural hydrogen accumulation. These stable, ancient continental areas often contain significant amounts of radiogenic materials, contributing to H_2 generation through radiolysis. Moreover, cratonic regions tend to have large, fractured zones where H_2 can accumulate in subsurface reservoirs. (iv) Sedimentary Basins: While H_2 generation predominantly occurs in crystalline or ultramafic rocks, sedimentary basins can act as traps for migrating H_2. Impermeable layers within these basins, such as shale or salt, may act as competent seals that inhibit the surfaceward migration of H_2. In some cases, H_2 may migrate into porous sandstone or carbonate reservoirs, where it can accumulate in commercial quantities. (v) Fault Zones and Tectonic Settings: Tectonically active areas, including rift zones and fault systems, provide pathways for H_2 to migrate from deep geological sources to shallower reservoirs. Faults create zones of high permeability where H_2 can move upward, often accumulating in overlying traps formed by impermeable rock layers or faults that create sealed compartments.

4.2 Natural Hydrogen System: From H_2 Molecule to Natural Hydrogen Reservoir

The formation of natural hydrogen arises from interdependent geological mechanisms encompassing H_2 generation, subsurface transport, accumulation, and confinement. These systems initiate with H_2 production via deep crustal processes such as serpentinization, radiolytic dissociation, or biogeochemical activity [12]. These H_2-generating processes are confined to distinct geological settings, including ultramafic lithologies and Precambrian basement complexes, which act as precursor zones for reservoir development. This foundational stage as the "Source Rock" unit, visualized as the primary locus of initial H_2 synthesis.

Following its generation, H_2 ascends through porous rock formations and fractures, which serve as natural conduits facilitating its upward transport. The pathways labeled as "Migration" in the image depict how H_2 is propelled by differential pressure gradients and the gas's intrinsic buoyancy as it traverses these geological features. H_2 is driven through these geological features by pressure differential gradients and inherent buoyant forces of the gas [13].

During upward migration, H_2 becomes trapped in porous geological formations, termed reservoirs, where accumulation occurs [12]. These reservoir units function as natural geological traps, facilitating the aggregation of H_2 into commercially viable quantities. Reservoir storage capacity is governed by parameters such as rock porosity and the dimensions of interconnected pore networks, which directly influence subsurface H_2 retention potential.

Effective trapping mechanisms are critical to prevent H_2 leakage from these reservoirs. Caprock is an impermeable rock layer that acts as a sealant, inhibiting further upward migration of H_2. This impermeable barrier is vital because hydrogen's low molecular weight renders it highly susceptible to escaping in the absence of an effective seal [14]. The structural configuration of traps, such as domal anticlines or fault systems, in combination with sealing units collaboratively ensures the retention of H_2 within the reservoir over geological time.

The H_2 system shares numerous similarities with petroleum and natural gas systems (Fig. 4.3). However, the formation mechanism of natural hydrogen fundamentally differs from hydrocarbon accumulations, exhibiting a dynamic reservoir-forming process: First, H_2 generation and consumption processes occur simultaneously in subsurface environments, where the genetic mechanisms and abundant source rocks of natural hydrogen ensure that the rate of H_2 generation exceeds that of consumption. Second, while H_2 escapes or is produced near the surface, continuous geological processes drive upward migration of deeper-sourced H_2 to recharge existing reservoirs or directly leak to the surface. This phenomenon has been demonstrated through long-term monitoring of H_2 microseepage in "fairy circle" structures across multiple nations and corresponding resource assessments. Once a H_2 reservoir forms and becomes exploitable, its instantaneous H_2 storage capacity and volume remain dynamically variable. The ultimate resource potential is principally constrained by the H_2-generating capacity of source rocks within the regional geological framework [15].

In summary, the dynamic interplay among these components, H_2 production, subsurface migration along permeable pathways, concentration within geological traps, and effective confinement, establishes a functional natural hydrogen system. Each phase of this sequence is interlinked, constituting an integrated framework that governs the feasibility of utilizing natural H_2 as a sustainable energy resource. The accompanying diagram offers a conceptual visualization of how geological mechanisms govern reservoir dynamics, highlighting the critical role of each element in preserving H_2 stability and retrievability within subsurface environments (reproduced with permission from [14]).

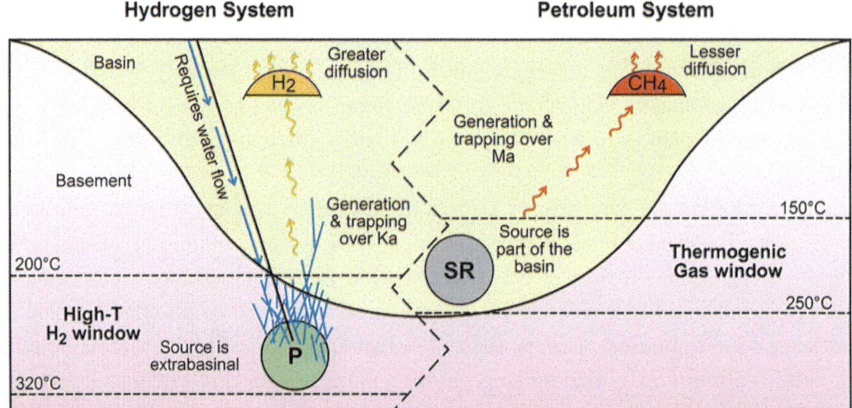

Fig. 4.3 Hydrogen system versus petroleum system. This comparison of fluid systems emphasizes the distinctions in the nature and spatial distribution of source rocks for reservoir-forming sediments. Note that SR = petroleum source rock; P = Protolith (reproduced with permission from [16])

4.3 Hydrogen Generation and Genesis Mechanism

Natural hydrogen is generated through various complex geological mechanisms within the Earth's crust. The genesis of H_2 is a complex and multifaceted process, thus not yet fully understood (Fig. 4.4). It is crucial to understand these processes to identify the potential H_2 reservoirs and effectively harness this clean energy source [1]. Several natural processes contribute to the generation of hydrogen on Earth [9].

Abiotic hydrogen: Abiotic H_2 is produced by geological and chemical processes without the involvement of biological organisms. Four key abiotic mechanisms contribute to the generation of natural hydrogen (Fig. 4.5), including: (i) Deep source degassing (e.g., deep magma and mantle). Deep-seated magmatic processes and mantle degassing represent the primary mechanisms for hydrogen generation, releasing H2-rich fluids from Earth's interior [17]. (ii) Serpentinization. Serpentinization, particularly within ultramafic rock formations, represents a key mechanism for natural hydrogen production via the oxidation of Fe(II)-rich minerals [18]. (iii) Water radiolysis. Water radiolysis, initiated by the radioactive decay of uranium, thorium, and potassium, serves as a contributing mechanism for the generation of hydrogen [2].

Biotic hydrogen (The organic origin of biological organisms): Biotic H_2 is produced through biological processes, typically by microorganisms in anoxic environments. Organic decomposition likewise constitutes a significant mechanism for hydrogen generation, wherein the microbial breakdown of organic material in anoxic environments yields H_2 [19]. These varied mechanisms reflect the diverse geological settings where natural hydrogen generation occurs, highlighting the imperative for further exploration

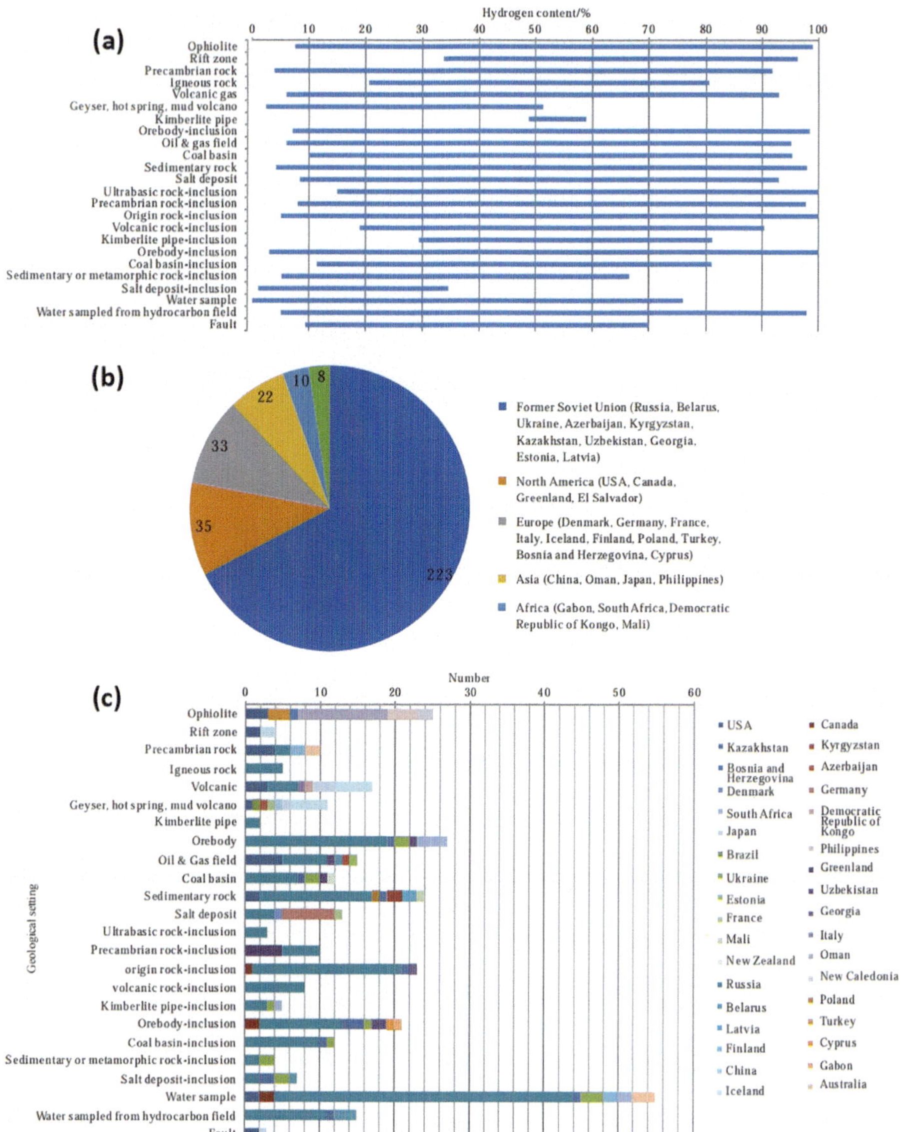

Fig. 4.4 Global discovery of natural hydrogen. **a** Measured hydrogen concentrations across various geological settings. **b** Regional distribution of documented hydrogen occurrences. **c** Spatial distribution of hydrogen occurrences categorized by both geological setting and geographic region [1, 23]

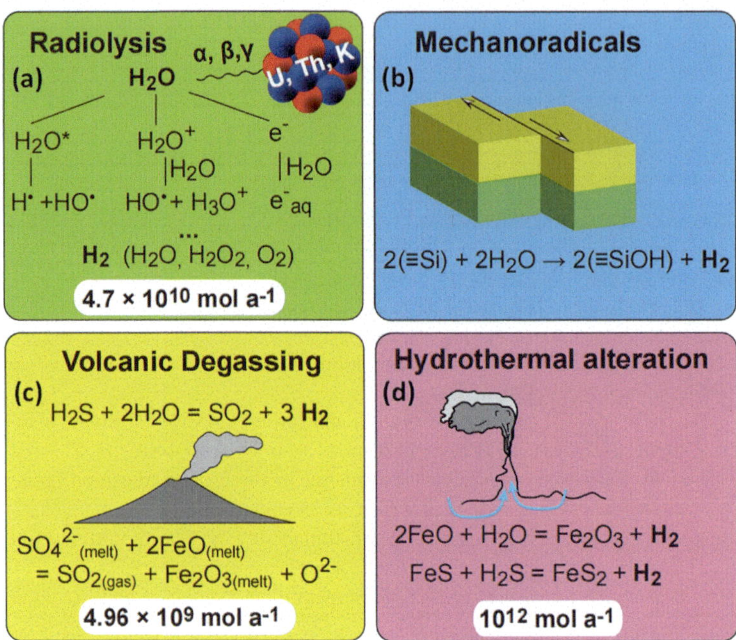

Fig. 4.5 Schematic diagram of the mechanisms of the genesis and origin of the abiotic hydrogen [21]. **a** Radiolysis produces H_2 during the dissociation of water through radioactive decay [24]. **b** Mechanoradical reactions generate H_2 on wet surfaces of active faults. **c** Volcanoes emit H_2 when SO_2 degasses at low pressure [25]. **d** During the hydration of rocks, H_2 is produced when water is reduced concurrently with the oxidation of Fe^{2+} to Fe^{3+} [26]

and understanding. For example, in the vicinity of the Bourakebougou deposit, the surrounding geological formations consist predominantly of Birrimian sequences, including plutonic, volcano-sedimentary, and sedimentary rocks. These are encircled by a suite of magmatic rocks that underwent deformation during the Eburnean orogeny, which occurred from approximately 2.2 to 2.0 Ga [20]. So, it can be inferred that hydrogen generation in this context is primarily attributed to magmatic degassing.

Among the major two natural processes (i.e., abiotic and biotic hydrogen generation), several primary geological processes have been involved including serpentinization, radiolysis, abiotic methanogenesis, and mantle degassing [11, 21, 22]. This book focuses primarily on the abiotic hydrogen.

4.3.1 Water–Rock Interactions in Crustal Fractures

H_2 can be produced via water–rock interaction processes occurring in a variety of crustal environments beyond ultramafic rocks, particularly via redox processes involving iron-bearing minerals [27]. In fractured crystalline lithologies such as granites and basalts, aqueous alteration of reduced iron-rich phases drives H_2 production through oxidation reactions, as documented in deep subsurface environments like continental boreholes and mine systems. Additionally, geothermally active regions with volcanic substrates facilitate elevated H_2 yields through high-temperature water–mineral interactions, where thermal dissociation of water molecules and Fe^{2+} oxidation in mafic assemblages synergistically enhance hydrogen generation [28].

The oxidative reaction can be represented a:

These reactions are often slow and less prolific than serpentinization, but they still contribute to the global natural hydrogen budget. Water–rock interaction encompasses the interplay between fluids and rocks occurring during various geological processes. The primary mechanisms for H_2 generation through water–rock interactions include serpentinization, reactions occurring between water and freshly exposed rock surfaces, and hydroxyl-mediated reactions within mineral structures. Among these, serpentinization represents the most extensively studied, as well as the most significant and prevalent mechanism for H_2 production through water–rock interactions [29].

4.3.1.1 Serpentinization

Serpentinization, a critical geochemical process in natural hydrogen formation, occurs when water interacts with ultramafic rocks rich in olivine [$(Mg, Fe)_2SiO_4$] and pyroxene, primarily at mid-ocean ridges, ophiolite complexes, and mantle rock exposures. This hydration reaction transforms olivine into serpentine minerals [$(Mg, Fe)_3Si_2O_5(OH)_4$] through mineralogical alteration, releasing molecular H_2 as a byproduct over extended geological timescales. The process is facilitated by hydrothermal circulation systems that enable deep crustal water penetration and sustained rock-fluid interactions, positioning serpentinization as a dominant mechanism for H_2 generation in reducing subsurface environments [30–32]. The key reactions involved in serpentinization is shown in Fig. 4.6.

Serpentinization represents a fundamental process driving substantial H_2 production and accounts for numerous observed H_2 seepages worldwide. This geochemical reaction is enabled by elevated temperatures and pressures (ranging from 200–500 °C), establishing ideal environments for H_2 liberation [13]. Such conditions predominantly occur in areas with active tectonic processes. Prominent geological settings exhibiting significant serpentinization activity encompass ophiolite formations and zones of oceanic crust, positioning these regions as prospective targets for natural hydrogen exploration (Fig. 4.7). Functioning as a persistent H_2 source, this process plays a crucial role in developing subterranean H_2 reservoirs. The H_2 produced through serpentinization migrates along rock fractures and fault systems, subsequently concentrating in permeable geological strata to

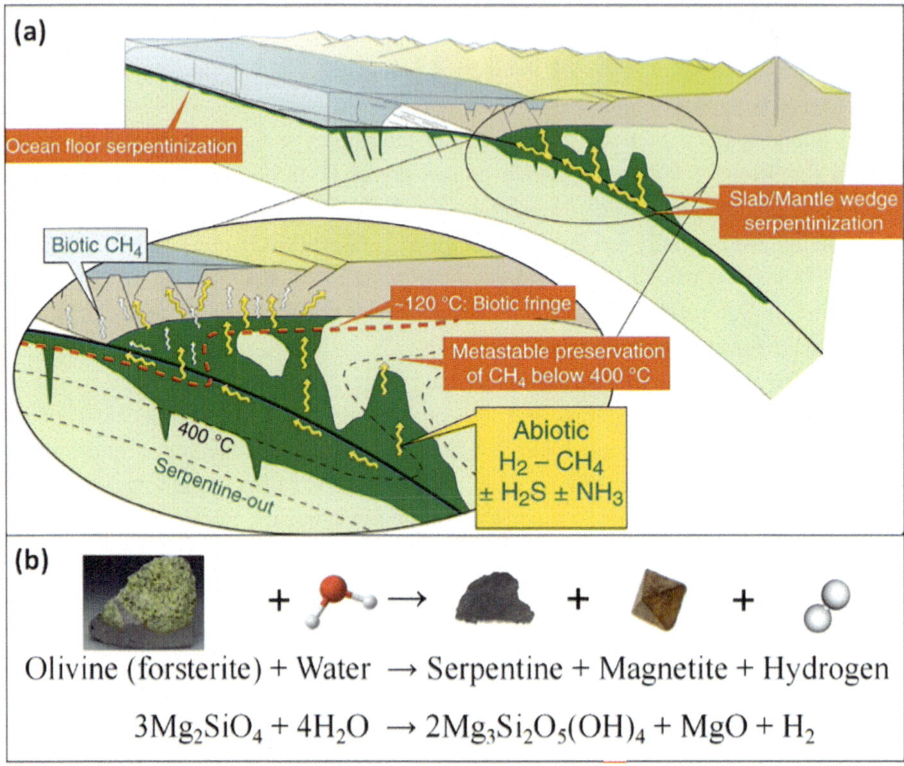

Fig. 4.6 Mechanism and chemical reactions of serpentinization. **a** Diagram of the serpentinization process. **b** The equation for the chemical reaction of serpentinization occurs in the reaction of water and rock (modified from [33])

form viable reservoir structures. This mechanism enables the natural storage of H_2 within subsurface traps through structural and stratigraphic containment.

Notable examples include the Semail Ophiolite in Oman, where H_2 seepage is associated with ongoing serpentinization reactions [34], and the Troodos Ophiolite in Cyprus [35]. In Western Australia's ultramafic rock formations, serpentinization occurs for large H_2 production potential [36].

Research indicates that serpentinization predominantly occurs within a temperature window of 200–310 °C, with optimal depths and geological settings determined by these thermal conditions [37]. Below this range, reaction kinetics are significantly hindered, whereas exceeding this threshold results in a sharp decline in reaction rates due to thermodynamic limitations. While earlier studies suggested H_2 production via low-temperature serpentinization, contemporary controlled experiments have revealed that detected H_2 in

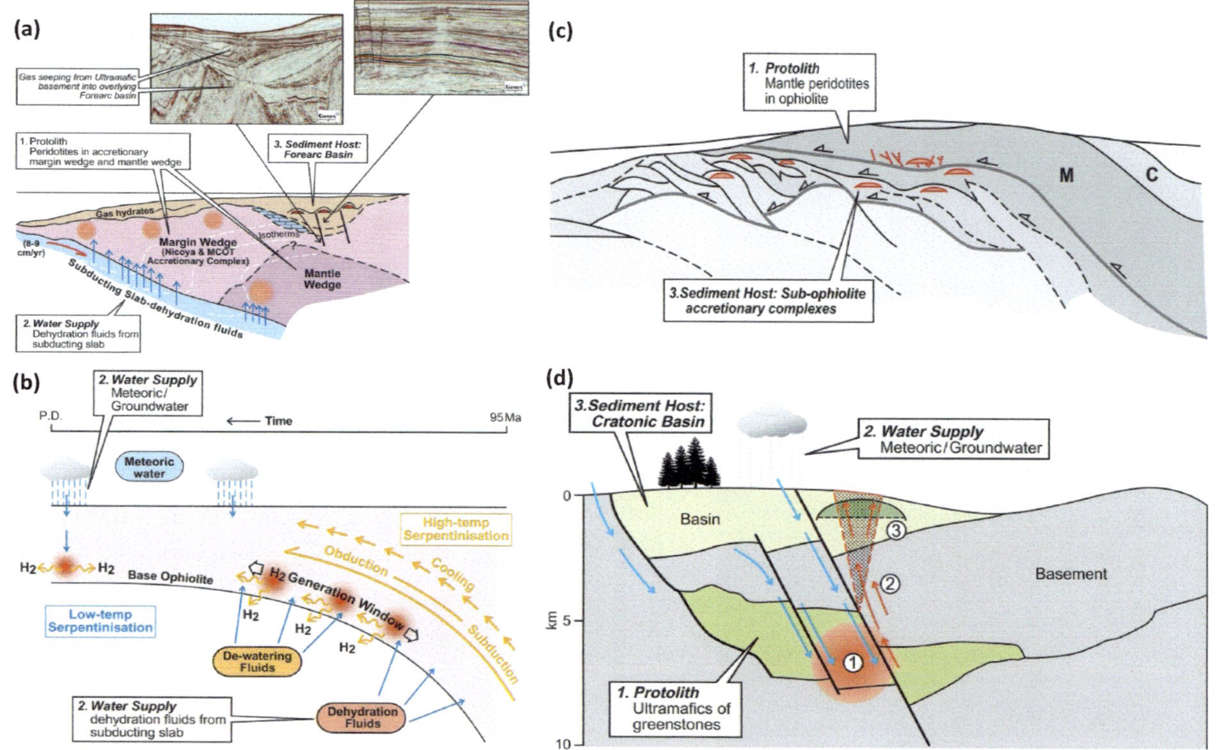

Fig. 4.7 Hydrogen production mechanism of ophiolite in different geological tectonic settings. **a** The peridotite of the accretion wedge and mantle wedge in the 'Cordillera' pre-arc basin reacts with the dehydration fluid of the subduction plate (Seismic images from Nicaragua courtesy of Geox MCG). **b** The early high-temperature and late low-temperature serpentinization processes of the "Tethys-type" ophiolite. **c** Omani ophiolite section. **d** The ultramafic rocks in the cratonolites react with surface precipitation, and the gas is transported to the sedimentary basin through faults to form hydrogen reservoirs (reproduced with permission from [4, 16, 54–56])

such cases stems from artifactual sources, specifically contamination from standard reaction vessels. These vessels, when analyzed in mineral-free control trials, were shown to intrinsically produce H_2, challenging prior claims of low-temperature H_2 generation [38].

It is also important to note that the serpentinization of peridotite leads to a reduction in permeability, thereby hindering further water extraction [39]. Significantly, some researchers propose that H_2 functions not only as a byproduct of serpentinization but also as a reactive participant in the formation of "structural" water incorporated into serpentine minerals [40]. This critical aspect requires further investigation to elucidate the complex hydrogeochemical processes involved in serpentinization reactions.

In regions where mantle material is exposed at the Earth's surface, serpentinization processes have been extensively studied. Mantle rocks approach the surface within rift zones, though these tectonic features predominantly occur beneath oceanic domains. Ophiolite belts, representing rare but more accessible exposures of uplifted oceanic crust and upper mantle sequences above sea level, provide critical natural laboratories for such investigations. Current estimates suggest that H_2 fluxes generated through serpentinization reactions in oceanic crustal systems range between $0.8–1.3 \times 10^{11}$ mol/year (0.16–0.26 Tg/yr) [41].

Previous research has hypothesized that serpentinized lithosphere constitutes approximately 10% of the seafloor, leading to a calculated maximum potential annual H_2 flux of 0.38 Tmol/yr 0.38 Tmol/yr (0.76 Tg/yr) [42]. Independent studies analyzing data from ocean crust drilling samples have estimated H_2 fluxes of $4.5 \pm 3.0 \times 10^{11}$ mol H2/yr (0.89 ± 0.6 Tg/yr) [43]. By extrapolating recent H_2 flux measurements from the Semail Ophiolite in Oman to the global distribution of ophiolites [44], some research yields a flux range of 0.18–0.36 Tg/yr, which is consistent with previous estimates in terms of order of magnitude. Furthermore, olivine has been identified as a promising candidate for sustainable H_2 production in energy systems.

The global estimated hydrogen flux emanating from mid-ocean ridge systems is 0.12 Tg/yr. This value appears relatively low compared to serpentinized peridotites. Other investigators have utilized $H_2/^3He$ and ^3He/heat ratios to quantify the global hydrogen flux emanating from slow-spreading ridges, deriving an estimated value of 89×10^9 mol/yr (0.18 Tg/yr) [45]. H_2 flux from slow-spreading ridges has been estimated at approximately 16.7×10^{10} mol/yr (0.33 Tg/yr) [46]. Subsequent studies provided a revised approximation of 19×10^{10} mol/yr (0.38 Tg/yr) [47]. More recent research incorporating diffusion rates into their models demonstrates significantly higher H_2 production, reaching magnitudes on the order of 10^{12} mol/yr [32].

In addition, molecular H_2 is primarily derived from processes associated with oceanic crust. A growing body of research proposes that hydrogen generation during the Precambrian may have been substantially underestimated. Previous estimates indicate that hydration processes, specifically water–rock interactions, may have contributed approximately $0.2–1.8 \times 10^{11}$ mol annually [48]. Mantle rocks are not the sole lithology capable of producing H_2 through water–rock interactions. Basalt-water reactions have also been demonstrated as a viable mechanism, representing another potential source rock for H_2 occurrence [49]. Investigations of basalt samples from oceanic crust obtained through the Integrated Ocean Drilling Program (IODP) revealed significant H_2 enrichment. The entire basaltic layer in the oceanic crust is estimated to produce approximately 17×10^9 mol of H_2 per day (12.6 Tg/yr) [25]. Alternative estimates suggest that global H_2 flux from basaltic oceanic crust could reach 7.5 Tg/yr [42]. Additional study indicates that H_2 generation through water interactions with Fe^{2+}-bearing minerals may occur at depths of 10–15 km. Microfractures with permeabilities exceeding 10^{-14} cm^2 can extend to 15 km depth, and such permeability are sufficient to permit significant fluid migration. It is estimated that basalt in aqueous equilibrium retains approximately 7 ppm (by weight) of H_2, corresponding to 75 cm^3 STP per kg of rock, or 12 L of H_2 per liter of rock-equilibrated water. However, some studies argue that basalt-groundwater interactions cannot produce significant H_2 quantities due to rapid passivation of water–mineral contact surfaces by reaction products [50]. Other Fe (II)-containing minerals have demonstrated H_2 generation potential through aqueous reactions. Research indicates that siderite ($FeCO_3$) can produce H_2 under specific conditions [51].

This issue about water consumption warrants detailed discussion, as most currently recognized geological sources of H_2 involve water–rock interactions (natural hydrogen is primarily attributed to serpentinization or alternative mechanisms detailed subsequently). Given the irreversible nature of these reactions, accurately determining water consumption rates is essential. With H_2 production rates under different geological mechanisms estimated at 23 ± 8 Tg/yr, assuming complete water consumption, the entire hydrosphere (1.4×10^{24} g) would be depleted within 6.6 ± 3 billion years. Notably, a portion of produced H_2 undergoes recycling into water through abiotic and/or biological consumption (via organic compound pathways), while oxygen remains permanently bound in rock matrices. Consequently, the H_2 generation process through water–rock reactions can be alternatively perceived as a redistribution mechanism of water's surface oxygen through lithospheric sequestration.

In natural environments, serpentinization reactions represent the primary mechanism for H_2 generation. This process fundamentally represents a mineral transformation wherein olivine and pyroxene minerals in mafic–ultramafic rocks are altered through hydrothermal metasomatism, resulting in the formation of various serpentine-group minerals. Such reactions predominantly occur at mid-ocean ridge systems, which currently serve as the most direct sites for crust-mantle interaction. These tectonic settings provide both migration pathways for mantle-derived fluids through mafic–ultramafic rock sequences and optimal thermal conditions (300–400 °C) required for serpentinization processes. Consequently, these environments facilitate extensive serpentinization reactions that generate H_2-rich fluids, with H_2 concentrations potentially exceeding 90% in the resulting gas phases.

In subduction zones, the presence of ophiolite remnants rich in olivine and pyroxene from oceanic crust leads to significant H_2 enrichment. Particularly, ophiolites that have not undergone high-pressure and low-temperature metamorphism retain olivine minerals either unaltered or with minimal serpentinization. During plate subduction processes, these ophiolites become tectonically intercalated within sedimentary sequences, occurring in either continental or oceanic crustal settings among clastic or carbonate rock formations. Major faults penetrating these ophiolite-bearing sedimentary strata facilitate groundwater circulation with varying salinity and pH conditions. This hydrogeochemical environment promotes serpentinization of olivine minerals, ultimately generating H_2-rich natural gas reservoirs through water–rock interactions [52].

The compositional variations of gas reservoirs in subduction zones are closely linked to the water supply conditions at different tectonic positions. When water availability is insufficient to meet the rate of serpentinization or maintains equilibrium with it, H_2 becomes the predominant gas, accounting for over 80% of the gas composition. Under conditions of abundant groundwater supply, dissolved CO_2 in water participates in subsequent reactions where previously generated H_2 reduces CO_2 to CH_4, resulting in a mixed gas reservoir containing both H_2 and methane. In scenarios with sufficient water availability and serpentinization occurring at shallower depths, dissolved atmospheric nitrogen combines with deep-sourced nitrogen to form nitrogen-rich gas reservoirs, where nitrogen concentrations may exceed 90% (Fig. 4.8) [52].

Serpentinization is facilitated by hydrothermal circulation processes within mafic–ultramafic rock assemblages, observed both at deep-sea hydrothermal vent systems and in continental ophiolitic sequences. The serpentinization process of fayalite releases H_2 ions (H^+). Generally, elevated magnesium concentrations in olivine and magnesium-rich silicate minerals tend to produce fluids with alkaline characteristics. Consequently, the hydration of ferromagnesian minerals, including olivine and pyroxene leads to the formation of serpentine, brucite, magnetite, and molecular H_2. The serpentinization reaction rate reaches its maximum within the temperature range of 200–310 °C, with diminished reaction rates observed both below and above this thermal threshold. Mantle rocks also undergo serpentinization. Under low-temperature conditions (< 100 °C), H_2 production is associated with spinel reactants and cubic crystal structures. The H_2 generation during serpentinization is primarily controlled by the abundance of ferrous iron in precursor minerals (e.g., olivine or pyroxene with differing Fe/Mg ratios) and the fraction that is oxidized to ferric iron (Fe^{3+}). A critical requirement for serpentinization is the presence of a strongly reducing environment, particularly at oxygen fugacities below those stipulated by the fayalite-magnetite-quartz (FMQ) buffer, concomitant with high hydrogen activity. Compared to basalt-water chemical interactions, serpentinization of ultramafic rocks typically produces higher H_2 concentrations. Serpentinization may occur predominantly in shallow crustal zones, specifically within the upper crustal depths of 4–6 km, which align with the 400 °C isotherm. However, the optimal temperature conditions for serpentinization reactions likely reside at greater depths of 10–12 km [29].

Furthermore, during low-temperature serpentinization, spinel minerals catalyze the oxidation of $Fe^{2+}$and subsequent H_2 production [27]. These mechanisms are classified into

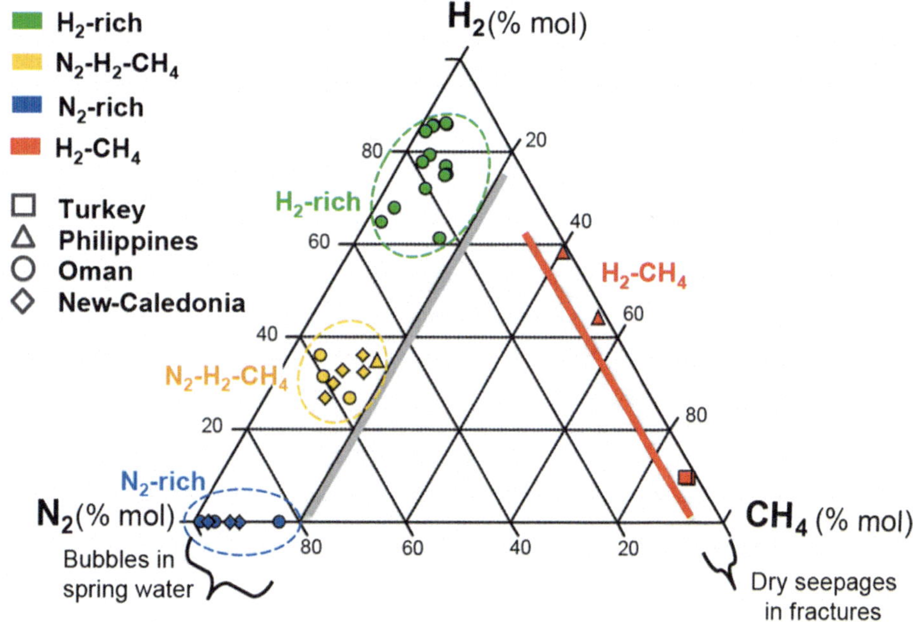

Fig. 4.8 The gas compositions analyzed from seeps in Turkey, Oman, New Caledonia, and the Philippines are plotted on a ternary diagram (mol%) representing the major components: H_2, N_2, and CH_4. Based on the relative proportions of these gases, four distinct types of gas mixtures can be identified. Each compositional type correlates with specific modes of seepage: N_2-bearing gases are associated with aquatic seeps, often discharging with water in streams, whereas N_2-free mixtures are linked to dry seeps emanating from fractures in massive rocks, which may spontaneously ignite. The aqueous seeps also exhibit distinct physicochemical characteristics, particularly in pH and temperature (reproduced with permission from [57])

two categories based on the Fe^{2+} source: (i) The first involves Fe^{2+} within the spinel mineral structure directly reducing adsorbed H_2O or H^+ on the mineral surface (Fig. 4.9a); (ii) The second entails Fe^{2+} ions, released from the dissolution of Fe^{2+}-rich minerals like olivine, adsorbing onto the spinel surface. Subsequently, electrons are transferred through the spinel from adsorbed Fe^{2+} to H_2O or H^+, resulting in the reduction of H_2O or H^+ and the formation of H_2 (Fig. 4.9b). Additionally, Ni^{2+} doping facilitates rapid phase transformation of $Fe(OH)_2$, yielding magnetite and generating H_2 (Fig. 4.10) [53].

4.3.1.2 Radiolysis

Radiolysis serves as a critical mechanism for H_2 generation in deep subsurface environments [1, 18, 21], where ionizing radiation from naturally occurring radioactive isotopes (uranium, thorium, and potassium) in host rocks drives water molecule dissociation into molecular H_2 and O_2. This process is facilitated by α-, β-, and γ-radiation emitted during radioactive decay, which disrupts water's molecular structure. Predominant in crystalline

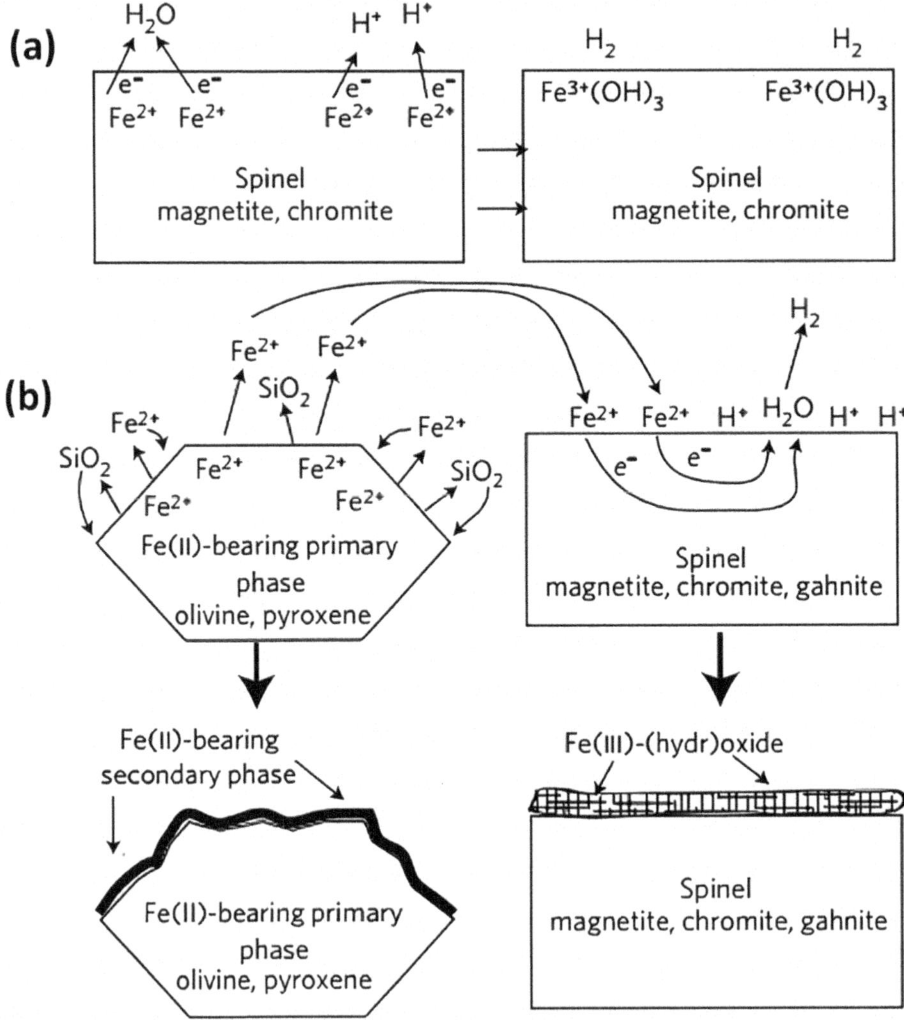

Fig. 4.9 Schematic illustration of spinel-induced hydrogen production during the interaction between olivine and water. **a** Fe (II) in the mineral lattice reduces the water molecules or hydrogen ions adsorbed on the surface. **b** Fe (II) released after the dissolution of other Fe(II)-containing minerals adsorbs to the surface of spinel minerals to reduce water molecules or hydrogen ions (reproduced with permission from [27])

basement lithologies such as granites and cratonic shields, radiolytic H_2 production correlates with the abundance of radioactive minerals in these settings. The generated O_2 is typically sequestered through mineral oxidation reactions, enabling H_2 accumulation in anoxic subsurface systems. The fundamental radiolytic reaction is commonly represented as:

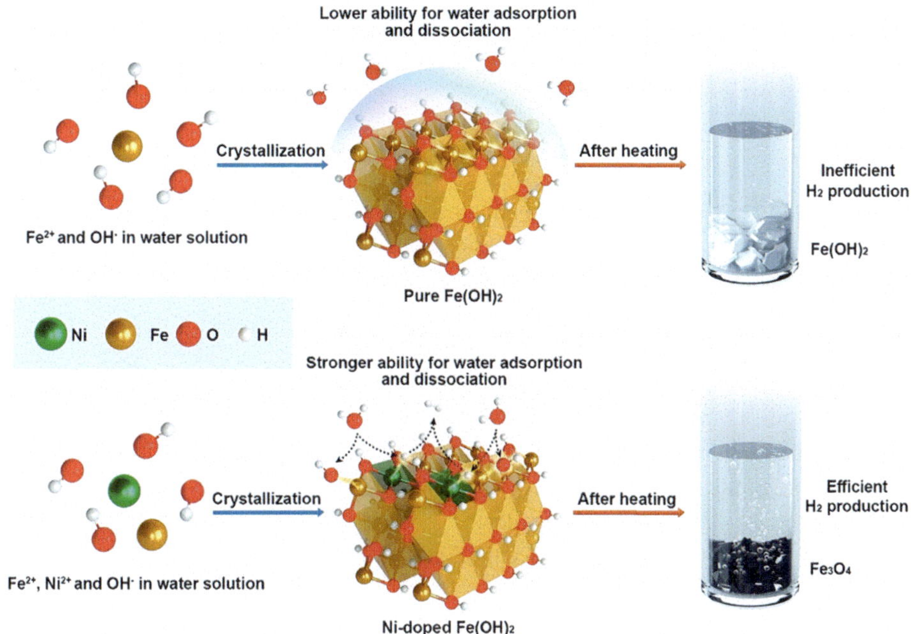

Fig. 4.10 The schematic illustration for Ni^{2+}-promoted H_2 production from $Fe(OH)_2$ oxidation during low-temperature serpentinization [53]

$$2H_2O \xrightarrow{radiation} 2H_2 + O_2$$

Radiolysis is a process that can take place across diverse lithologies (Fig. 4.11), yet it gains particular significance in ancient geological settings like Precambrian shields, which host elevated concentrations of radioactive minerals [58]. This phenomenon predominantly unfolds in ancient cratonic rocks, which form stable continental cores termed cratons. Notable cratonic regions, including the Canadian Shield, Scandinavian Shield, South Africa's Witwatersrand Basin, and analogous ancient terrains, exhibit elevated uranium concentrations that facilitate continuous H_2 generation via radiolysis [1]. Consequently, these areas represent optimal sites for sustained radiolytic H_2 production. The radiolytic mechanism is pivotal for H_2 formation in deep subsurface systems, where uninterrupted radioactive decay persists over geological timescales. Generated H_2 can migrate through geological strata and become trapped in fractures or structural reservoirs, creating economically significant natural hydrogen accumulations. These cratonic environments provide enduring radioactive sources, ensuring H_2 production across millions of years.

Water radiolysis is well-established as a significant process contributing to the generation of H_2 in geological systems. The radiolytic dissociation of water into H_2 and O_2 is driven by energy derived from radioactive decay. This process is particularly relevant in

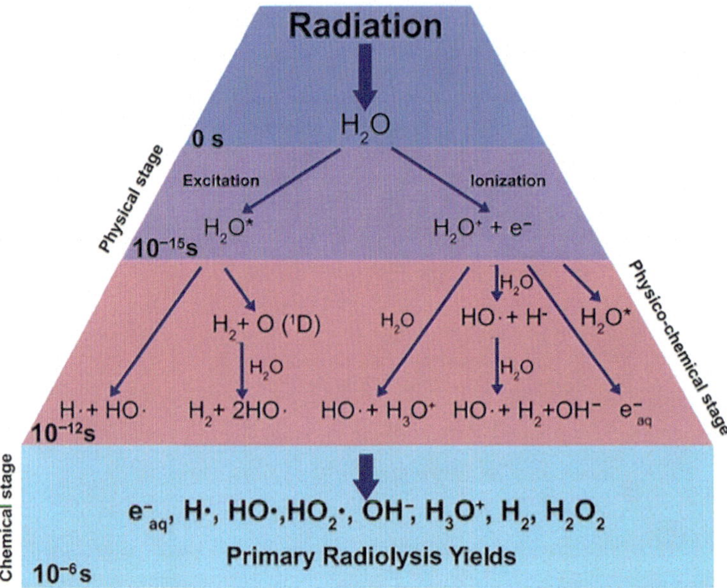

Fig. 4.11 Diagram of radiolysis processes (reproduced with permission from [59])

geologic contexts due to the ubiquitous presence of naturally radioactive elements such as uranium (U), thorium (Th), and potassium (K) within the Earth's crust. The hypothesis of mineralogical transformations driven by ionizing radiation was originally introduced by V. I. Vernadsky in the 1930s. Notably, some studies argue that existing alternative H_2-generation mechanisms cannot adequately explain the large-scale production of H_2 observed in certain geological settings, underscoring radiolysis as a critical process [2].

Ionizing radiation causes molecular excitation and ionization, which increases chemical reactivity and promotes the generation of free radicals (highly reactive chemical species). For instance, a single 1 MeV α-particle can ionize up to 105 molecules during its energy dissipation process [60]. When this energy transfer occurs in aqueous systems, water molecules undergo radiolytic decomposition to produce H_2 and H_2O_2, with the latter rapidly dissociating into O_2 and water. Current models indicate that only an estimated 1% of the total energy generated from radioactive decay is absorbed by pore fluids, with the vast majority being assimilated by the mineral matrix and dissipated as heat. Consequently, H_2 production rates exhibit direct proportionality to the water-filled porosity of host rocks [2]. A specialized monograph addressing radiolysis in geochemical systems presents more conservative values: merely 0.002–0.06% of radioactive energy in igneous rocks and 0.03–0.6% in clay-rich formations contributes to chemical potential transformation of compounds, with the majority dissipated thermally. In multicomponent geological systems, radiation energy distributes among all constituent phases, resulting in only partial energy allocation for aqueous radiolysis [61]. Recent experimental investigations demonstrate that γ-irradiation of H_2-bearing molecules and mixtures (including aqueous solutions) generates measurable H_2. Notably, brine systems exhibit enhanced H_2 yields compared to pure water. Nevertheless, some studies have proposed mechanistic frameworks for hydrogen generation via radiolysis, despite the absence of direct quantitative measurements of H_2 production [62].

Estimates of H_2 generation from these processes exhibit significant variability. One author suggests that the volume of water decomposed by radiolysis in sedimentary profiles is approximately 0.026 m^3/yr (which may represent a typographical error, as the author likely intended km^3; even then, this value remains relatively low: only 0.00032 Tg H_2/yr) [63]. A monograph dedicated to water radiolysis presents significantly higher values: 0.24×10^{23} g over 4×10^9 years (equivalent to 6 Tg/yr) [61]. Recent evaluations of radiolytic processes within continental Precambrian basement rocks suggest hydrogen production rates between 0.16 and 0.47×10^{11} mol/yr. [26].

It is essential to emphasize that H_2 represents only one of multiple products formed during water radiolysis. The transformation of radioactive decay energy within water/rock systems leads to the co-generation of both oxidizing and reducing species, such as H_2, which exhibit contrasting chemical behaviors [63]. Under aqueous decomposition conditions, this mechanism predominantly generates H_2O_2, a compound that rapidly decomposes into molecular oxygen (O_2). Laboratory irradiation experiments involving water/mineral mixtures containing common rock-forming minerals have yielded hydrogen–oxygen gas blends with oxygen contents reaching 30–35% [61]. Despite this, environments where water radiolysis is hypothesized to operate do not exhibit the near-stoichiometrically balanced H_2/O_2 ratios predicted by such processes. For instance, oxygen was conspicuously absent in fluid inclusion analyses [64]. Intriguingly, Raman spectroscopy studies of quartz fluid inclusions from the Oklo Precambrian uranium deposit (Gabon) detected exclusively H_2, whereas inclusions from two Canadian Precambrian uranium deposits contained almost pure oxygen [60]. These anomalous gas compositions, complete oxygen depletion in Oklo samples and H_2 absence in Canadian examples, cannot be reconciled with radiolytic mechanisms, as they fundamentally violate the stoichiometric principles governing water radiolysis-derived gas formation.

In a separate investigation, H_2 production rates were estimated by analyzing the total groundwater residence time in aquifers, projected to span 3 to 80 million years. A comparison between modeled H_2 yields and empirically measured H_2 concentrations in natural systems revealed values of comparable magnitudes [2]. However, hydrogen's high mobility as a gas underscores the need to account for its diffusion rate, a critical factor omitted in these earlier calculations. Studies of fluid inclusions in rocks of diverse ages revealed no systematic relationship between H_2 abundance and geological age [65]. Conversely, a contrasting study reported elevated H_2 levels in Precambrian rock samples relative to younger geological units [64].

When estimating H_2 generation over geological timescales, the diffusion of H_2 must be taken into consideration. As H_2 continuously migrates away from reaction zones, the oxidation state of residual fluids is expected to increase [60]. Radiogenic decomposition of water has been estimated to produce 1.91×10^{23} g of oxygen in the Earth's crust over 4×10^9 years, a quantity sufficient to oxidize the entire crustal reservoir [61]. A substantial body of literature that invokes water radiolysis as a plausible mechanism for natural hydrogen generation has largely overlooked the corresponding accumulation of oxygen, thereby neglecting oxidation pathways associated with radiolytic production of oxidizing species. No detectable hydrogen peroxide was identified within rock samples exposed to ionizing radiation [2]. Other hyperoxidized compounds in crustal environments remain exceptionally rare, with documented occurrences limited to two uranium peroxide minerals [66].

An alternative H_2 production mechanism linked to elemental radioactive decay arises from valence state alterations. During radioactive decay, isotopes are transformed into daughter elements that exhibit distinct chemical properties. These daughter products display enhanced reactivity relative to their parent isotopes, with their collective reactivity surpassing half that of the original parent atoms. This amplified reactivity is predominantly attributed to the buildup of radiogenic ^{40}Ca and ^{87}Sr, formed via the decay of ^{40}K and ^{87}Rb, respectively. These secondary elements participate in aqueous reactions that release H_2. Additionally, the decay of Th and U into Pb liberates surplus oxygen atoms, elevating the oxidation potential of the surrounding environment. The authors posit that the H_2 yield from this process is analogous in magnitude to radiogenic helium production from thorium and uranium decay. Within a 100 km^3 volume of crustal rock, the calculated annual hydrogen production via this mechanism is approximately 0.15 m^3 [67].

One specific case merits closer examination. Consistently elevated H_2 concentrations have been documented in potash deposits. The prevailing hypothesis attributes this H_2 to water radiolysis triggered by radioactive decay of potassium (K) and rubidium (Rb) [64, 68]. In carnallite and sylvite, radiation emitted by ^{40}K interacts with water molecules, generating H_2 that accumulates in pore spaces, occasionally attaining pressures reaching megapascal levels [63]. Simultaneously, the oxidizing species generated through this process oxidize iron to hematite, which contributes to the distinctive red hue observed in these salts [67]. Nevertheless, measured H_2/Ar ratios and comparative H_2 concentrations in carnallite versus sylvite contradict this proposed mechanism [68]. Additional experimental evidence questioning the radiolytic origin of H_2 in potassium- and sodium-bearing minerals stems from irradiation studies: notably, experiments exposing cyanochroite to ionizing radiation failed to detect H_2 production [69].

An alternative origin for hydrogen generation within saline formations has been postulated. Notably, studies have reported that intact salt blocks release greater quantities of H_2 upon dissolution in water relative to fragmented salt material [68, 70]. In potassium-rich deposits, the branched decay of radioactive ^{40}K yields ^{40}Ar (12%) and ^{40}Ca (88%)

as daughter products. While a portion of ^{40}Ca may form chloride compounds, residual calcium is retained in its elemental form within salt mineral lattices. When exposed to aqueous environments during mineral dissolution, this metallic calcium undergoes hydrolysis, producing molecular H_2 [70]. Supplementary research indicates that H_2 production may also involve ^{87}Sr, a decay product of ^{87}Rb, reacting with water in addition to ^{40}Ca. However, inconsistencies between theoretically predicted and empirically measured H_2 yields have been documented [67]. Independent investigations challenge the exclusivity of the ^{40}Ca-water interaction as the sole H_2 source, positing complementary contributions from radiation-induced water dissociation [61]. Furthermore, carnallite dissolution processes may facilitate secondary chemical reactions that release occluded gases. A proposed mechanism suggests H_2 formation via redox interactions between radiation-generated F-centers in carnallite crystals and aqueous solutions [68].

Consideration should also be given to other processes associated with water radiolysis. Inert gases are generated as radiogenic products from the radioactive disintegration of certain isotopic elements. Helium-4 (^4He) is generated through radioactive decay chains of U and Th, while the radioactive decay of ^{40} K produces ^{40}Ar. These gases should co-occur with H_2 if water radiolysis is associated with the decay of these elements. Given the aforementioned radioactive conversion rates to H_2 ($\leq 1\%$), the quantity of ^4He should substantially exceed that of H_2. In fact, existing studies have documented that radiolytic H_2 production from water cannot surpass the amount of radiogenic helium [67]. Furthermore, any estimation of radiolytic H_2 generation must account for the recombination rates of oxidizing and reducing species formed through water interaction with high-energy particles.

Therefore, the radiolysis of water never constitutes an isolated reaction, but rather a complex of interconnected processes. In studies emphasizing H_2 generation, researchers often fail to account for the concurrent production of oxidizing species through radiolytic decomposition. When substantial volumes of H_2 are present, it is unlikely that the origin can be attributed exclusively to radiolytic decomposition of water. In fact, studies have concluded that radioactive decay alone cannot account for the observed volumes of molecular hydrogen (radiolytic hydrocarbon generation constitutes only a minor fraction) [71].

Aqueous radiolysis within lithospheric environments constitutes another important process for the abiotic production of hydrogen. The lithosphere hosts significant quantities of radioactive elements, including uranium, thorium, and potassium, whose decay produces α, β, and γ radiation. This energy drives the dissociation of water molecules into O_2 and H_2. Notably, saline water demonstrates enhanced H_2 yield compared to pure water systems. During aqueous radiolysis, hydrogen peroxide should theoretically emerge as the primary oxidant, which subsequently decomposes to produce oxygen. This process would predict oxygen concentrations reaching 30%–35%, a projection that contradicts actual observations. Consequently, water radiolysis constitutes not a singular reaction but rather an intricate network of interdependent chemical processes. While H_2 generation occurs, it is inevitably accompanied by the production of oxidizing chemical species. When exceptionally large H_2 quantities are detected, they cannot be attributed solely to water decomposition induced by radioactive decay [29].

4.3.1.3 Silicon Radical-Induced Reduction of H_2O Molecules to Produce H_2

The fragmentation of silicon (Si)-bearing minerals generates dangling bond-silica radicals (Si) on freshly fractured mineral surfaces, a process driven by the distinct chemical characteristics of Si–O bonds. These radicals facilitate the reduction of water molecules, yielding H_2. Because the formation of Si· radicals relies on ongoing mechanical fracturing to expose fresh mineral surfaces, this H_2-generating mechanism is strongly linked to tectonic fault activity. For instance, pronounced H_2 emissions have been documented in seismically active zones, such as the northern Yilan-Itong fault in China and the west-central Heilongjiang fault region. Studies indicate that surface concentrations of Si radicals increase markedly during intense mechanical disaggregation of minerals, enabling sustained reductive reactions with H_2O to yield H_2 (Fig. 4.12) [72]. Despite the widespread occurrence of Si-mediated H_2 production on fractured rock surfaces globally, no comprehensive global-scale quantification of this process has yet been undertaken [53].

4.3.2 Mantle Magmatic Degassing (Deep-Seated)

H_2 is additionally liberated from the Earth's mantle via magmatic degassing processes. This phenomenon involves the exudation of volatile compounds, such as H_2, from the mantle through volcanic and tectonic mechanisms [73–76]. Such degassing transpires via both explosive volcanic events and continuous venting along mid-oceanic ridge systems and tectonic plate boundaries.

Mantle degassing plays a particularly important role in volcanic environments, observed at mid-ocean ridges, subduction zones, and other tectonically active regions where mantle-derived materials rise toward the surface. In these settings, degassing processes in mantle-derived rocks liberate considerable amounts of H_2 (Fig. 4.13), alongside other volcanic gases (notably carbon dioxide and methane) [12]. For instance, during magma ascent toward the surface, volatile components (including H_2) are progressively released from the molten material. The predominant volatile species emitted during volcanic eruptions are predominantly water vapor (H_2O), carbon dioxide (CO_2), and sulfur compounds, H_2 can constitute a notable fraction of gas outputs in certain magmatic systems, especially those associated with mafic and ultramafic lithologies. This degassing mechanism occurs more frequently in: (i) Volcanic regions: H_2 is recognized as a constituent of volcanic gas emissions. Regions characterized by active volcanism, such as the East African Rift System, exhibit optimal conditions for mantle degassing processes. Notably, the Rungwe Volcanic Province in Tanzania has been documented as a site where

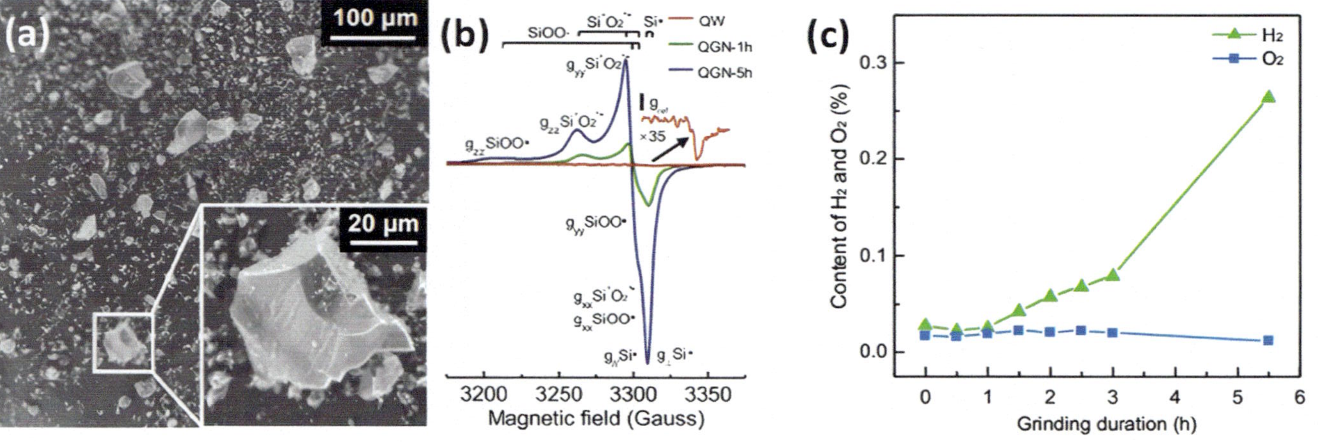

Fig. 4.12 Microscopic characterization, elemental distribution, and hydrogen production activity of ground quartz. **a** SEM image. **b** The detection of Si. **c** Hydrogen production after quartz grinding under N_2 atmosphere for 5 h (adapted from [53, 72])

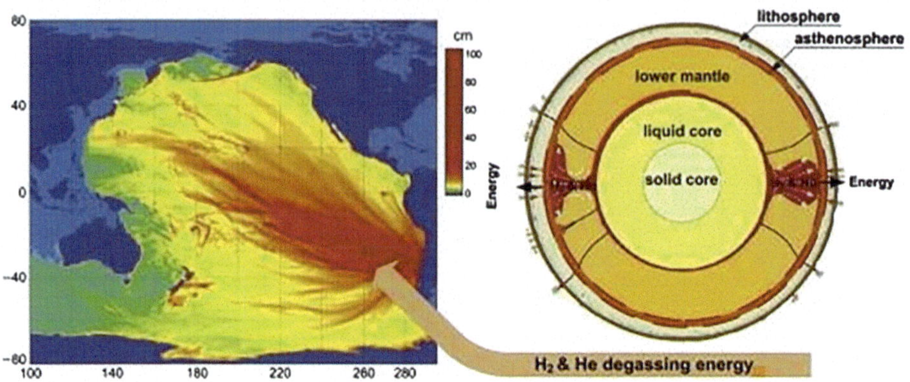

Fig. 4.13 Degassing from the outer mantle [79]

mantle degassing drives substantial natural hydrogen emissions [77]. Persistent tectonic activity, combined with elevated geothermal gradients in these areas, promotes the generation and sequestration of H_2 reservoirs. (ii) Geothermal systems: H_2 may be present within hot magmatic fluids and concentrate within fractures or fault systems. Regions exhibiting active geothermal processes, such as the East African Rift and Iceland, represent prominent candidates for the generation and release of mantle-derived H_2 [1].

H_2 can migrate through subsurface pathways and become concentrated in fractures and permeable geological strata [78]. When effective trapping systems, such as impermeable caprocks or structural closures, are present, H_2 may be preserved in economically significant accumulations within subsurface reservoirs [18]. While magmatic degassing represents a less prevalent mechanism for H_2 generation compared to serpentinization, it remains a notable contributor to natural hydrogen emissions in volcanic and geothermal environments, such as Iceland, as well as regions characterized by active tectonic processes.

Deep-source H_2 can be categorized into primordial and secondary H_2 based on its origin. Primordial H_2 primarily refers to H_2 gradually degassed from the mantle or core to the surface, in contrast, secondary hydrogen is produced via chemical reactions occurring in the mantle and crust [1]. The majority of research has observed that hydrogen abundance increases proportionally with depth in ultra-deep boreholes. However, since current surface drilling cannot reach the mantle or core, this observation supports the hypothesis of mantle/core degassing as a H_2 source [80, 81]. Notably, evidence of H_2 presence has been identified in metallic inclusions within mantle-derived superdeep diamonds [76]. Experimental studies demonstrate that H_2 can dissolve in minerals under mantle-reducing conditions [82], further corroborating mantle/core degassing mechanisms. Nevertheless, some researchers remain cautious about deep degassing theories due to the requirement of faults or fracture zones for gas accumulation and surface migration [83].

At present, it is widely accepted that the lower mantle environment contains abundant CH_4, H_2O, and H_2. The redox gradients prevalent in mantle conditions create a fluid distribution spectrum ranging from water-dominated to H_2-rich compositions [84]. Current hypotheses suggest this phenomenon may be associated with H_2 migration from the Earth's core. Furthermore, substantial indirect evidence supports the potential H_2 enrichment in both mantle and core reservoirs. Notably, the long-standing paradox of the "core density deficit", which refers to the discrepancy wherein the Earth's core is estimated to be about 10% less dense than expected for a pure Fe–Ni alloy at corresponding pressure–temperature conditions, may be resolved by the incorporation of hydrogen [85, 86]. Experimental data demonstrate that $FeH_{0.14}$ in the inner core and $Fe_{0.88}Si_{0.12}H_{0.17}$ in the outer core exhibit density profiles consistent with seismic observations [85, 87]. The relative stability of iron hydrides in mantle-core environments supports the hypothesis that H_2 content equivalent to ~ 1 wt% could effectively reconcile the observed density discrepancy [86].

In addition, studies on the distribution of ionization potentials of elements in the solar system indirectly indicate that Earth's composition is rich in H_2 [88]. Based on models of solar system and Earth formation and evolution, large quantities of methane and other non-hydrocarbon resources exist in Earth's deep interior. These methane reserves were present since Earth's formation, with substantial amounts of reduced organic carbon being released through heating processes in the deep crust. Through various geological processes over time, this methane migrated upward and accumulated extensively at depths around 15 km in the crust, forming abiotic oil and gas reservoirs [89]. According to Earth evolution theory, the atmosphere formed approximately 3500 Ma ago through mantle degassing was reducing in nature, containing methane and other gases but lacking free oxygen. During Earth's accretion phase, primordial atmospheric methane was preserved as "fossil" gases in the upper mantle and deep crust. The release of these methane reserves occurs through tectonic fractures, volcanic activities, and major orogenic events (e.g., Caledonian and Hercynian movements), particularly in the presence of superplumes and deep-seated faults (rifts). This emission process has persisted throughout geological history with several major release episodes. While most mantle-derived gases dissipate into the atmosphere, a fraction accumulates as natural gas reservoirs. Currently, such reservoirs are predominantly associated with sedimentary strata, though volcanic rock-hosted reservoirs are being increasingly discovered. Notably, deep-sourced gases may mix with biogenic hydrocarbons [90]. Examples of such methane occurrences have been documented in the East Pacific Rise, Red Sea, Iceland, and volcanic regions including Wudalianchi (Northeast China) and Tengchong (Southwest China). Gold suggested that hydrocarbon prospects are particularly promising in continental plate marginal fold belts, major crustal rifts, seismic zones, volcanic regions (active or extinct), and linear extensions of known hydrocarbon-bearing trends. As previously discussed, mantle-derived hydrocarbons can enter the atmosphere, migrate into sedimentary reservoirs, or become incorporated into igneous/metamorphic rocks and hydrospheric systems. The discovery of extensive gas hydrates in Arctic regions represents methane migration products forming ice-like compounds. Gold further analyzed the mechanical processes governing the upward migration of deep methane.

The Geological Institute of the former Soviet Academy of Sciences attached great importance to the study of deep-seated gases [91]. Based on their theoretical framework, experimental simulations, and extensive geochemical data, they demonstrated that H_2, various hydrocarbon gases, and H_2S are formed under strongly reducing conditions in deep Earth sources. Their study suggested that within the specific pressure and temperature regime characteristic of the upper mantle, the liquid–gas phase serves as a massive reservoir for H_2 and hydrocarbons. The degassing processes of H_2 and methane migrating from deep Earth sources to the crust's surface are tectonically controlled. Structural features such as deep fault zones, rift valleys, grabens, aulacogens (failed rifts), and flexure-fault zones act as vertical migration pathways for deep-seated gases, forming what they termed "degassing conduits". Thermodynamic calculations under conditions of 65 kPa pressure and 1700 °C revealed that high pressure not only inhibits hydrocarbon thermal decomposition but also promotes hydrocarbon cyclization, polymerization, and condensation processes, facilitating the evolution towards higher molecular weight hydrocarbons. This provides experimental evidence for deep-source gas-to-oil synthesis. This deep-source gas theory has been practically applied to ultra-deep well exploration. The former Soviet Union implemented an ultra-deep drilling program across 11 regions to verify deep Earth gas theories. Five ultra-deep wells were operational, with the SG-3 borehole on the Kola Peninsula of the Baltic Shield being particularly notable. Commencing in May 1975, it reached 12,066 m by December 1983, becoming the world's deepest well. At 7000 m depth within the Archean Kola Group's gneisses and amphibolites, the well encountered asphalt inclusions accompanied by high concentrations of H_2, methane, helium, nitrogen, and brine. These results constitute conclusive evidence for the presence of abiogenic methane and hydrogen in deep crustal settings.

An integrated assessment of fluid evolution across Earth's major reservoirs demonstrates that fluids derived from the mantle exhibit more reduced redox conditions compared to those originating from the crust. Under such reducing environments, significant amounts of H_2 could potentially be retained within the mantle. Furthermore, studies demonstrate that the upper mantle and major asthenosphere of Earth have achieved metallic saturation, wherein the predominant fluid phase should be H_2-enriched. H_2 isotope investigations reveal that the mantle contains molecular H_2 and serves as a H_2 supplier to crustal rocks, particularly "surface" rocks. Recent relevant studies suggest that substantial H_2 might be stored in the mantle as hydrated mineral phases. A groundbreaking discovery by researchers, the first identification of natural vanadium hydride (VH_2) occurrence, offers conclusive evidence for the presence of H_2-rich fluid phases in the mantle [29].

4.3.3 Chemical Weathering and Oxidation of Reduced Gases

In some cases, the weathering of reduced minerals, particularly iron-bearing minerals in both ultramafic and mafic rocks, can release H_2. The oxidation of these reduced minerals by surface or groundwater leads to H_2 release. This process typically occurs in highly weathered or oxidized environments and is slower than other mechanisms, but it can still contribute to small amounts of H_2 generation [1].

4.3.4 Microbial Activity

Although abiotic processes dominate current research on natural hydrogen formation, microbial activities also represent a noteworthy source of H_2 production (Fig. 4.14). Specifically, H_2 can be generated through microbial mechanisms such as (i) fermentation and (ii) photosynthesis. These processes have been shown to significantly contribute to H_2 generation in both natural and engineered systems [78]. In anaerobic (oxygen-poor) environments, certain microorganisms metabolize organic matter (OM) via fermentation, releasing H_2 as a metabolic byproduct. For example, bacteria such as Clostridium species generate H_2 during the anaerobic digestion of carbohydrates. The chemical equation below illustrates the metabolic pathway by which these bacteria convert glucose into H_2.

$$C_6H_{12}O_6 \rightarrow 2C_2H_5OH + 2CO_2 + 4H_2$$

Within sedimentary basins, microbial processes and organic matter breakdown can generate H_2 through anaerobic metabolic activity. This biogenic H_2 production occurs when microorganisms decompose organic substrates in oxygen-depleted environments, potentially creating localized accumulations within suitable geological formations [11]. While

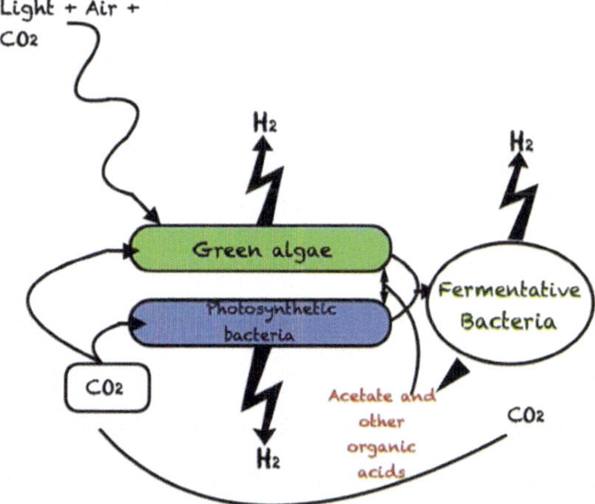

Fig. 4.14 Microbial activity producing hydrogen

this mechanism contributes to H_2 enrichment in organic-rich sedimentary basins its global significance remains comparatively lower than abiotic processes such as serpentinization or radiolysis.

The origin of natural hydrogen in gas reservoirs is often attributed to biogenic processes, involving H_2 generation through anaerobic decomposition of organic matter, fermentation processes, and metabolic activities of nitrogen-fixing bacteria. In natural settings, microbial communities that generate H_2 coexist with those that consume it, resulting in the rapid assimilation of biologically produced hydrogen into other chemical species. Two principal mechanisms govern H_2 decomposition: microbial oxidation and abiotic soil-mediated processes. Soil systems exhibit negligible H_2 generation rates compared to consumption rates. In wetland soils, while H_2 production occurs rapidly, instantaneous conversion to methane by methanogenic archaea prevents accumulation. Arid soil environments, characterized by aerobic conditions, demonstrate H_2 generation primarily through nitrogen-fixing bacteria, followed by abiotic enzymatic oxidation. The contribution of H_2 from organic matter fermentation remains relatively minor. During sedimentary metamorphism, the aromatization of residual kerogen progressively intensifies, yielding substantial methane quantities accompanied by subordinate H_2 production. Thermal cracking of alkanes and methane takes place at high temperatures (exceeding 200 °C and 500 °C, respectively), leading to the generation of H_2. However, under natural thermal conditions, generated H_2 undergoes immediate reaction with oxygen-bearing compounds, preferentially forming thermodynamically stable water through oxidation processes [29].

4.3.4.1 Fermentation

Select anaerobic microorganisms, including members of the *Clostridium* spp., generate H_2 as a metabolic byproduct during the fermentation of organic substrates. This occurs when these microbes catabolize carbohydrate-rich substrates under anaerobic conditions, a process that results in the release of molecular H_2.

$$C_6H_{12}O_6 \rightarrow 2C_2H_5OH + 2CO_2 + 4H_2$$

Fermentation commonly takes place in organic-rich settings such as sedimentary basins, wetlands, and peat bogs, which are characterized by abundant organic substrates [78]. Such environments favor the anoxic conditions necessary for optimal microbial metabolism and H_2 generation.

4.3.4.2 Photosynthesis

Photosynthetic microorganisms, including cyanobacteria and green algae, possess the ability to produce H_2 through a light-dependent process [78]. These organisms harness solar energy to split water molecules under anaerobic conditions [20], resulting in the generation of H_2 and O_2:

$$2H_2O \rightarrow 2H_2 + O_2$$

H_2 production driven by photosynthetic processes predominantly occurs in shallow aquatic ecosystems, where sunlight penetration facilitates microbial photosynthesis and sustains microbial communities. In sedimentary basins abundant in organic matter (OM), microbial-mediated H_2 generation is especially pronounced, as the breakdown of OM by microorganisms can yield localized H_2 accumulations. Under certain environmental constraints, such as nutrient scarcity or exposure to distinct light spectra, these microbial communities may adjust their metabolic pathways to favor H_2 generation. A well-documented case is the Appalachian coal basins in the United States, where H_2 emissions linked to microbial activity have been documented in these geologic settings [92].

In conclusion, while abiotic processes remain the dominant mechanism for natural hydrogen generation, microbial processes under specific conditions also contribute to H_2 production. Within anoxic settings, including deep marine sedimentary systems, certain prokaryotic organisms, namely bacteria and archaea, produce H_2 as a byproduct of metabolic processes. Although microbially generated hydrogen can accumulate in porous reservoir rocks, fracture systems, and structural traps, forming exploitable accumulations, such biological processes are not typically responsible for significant natural hydrogen reserves. Harnessing these microbial mechanisms presents a promising avenue for sustainable H_2 production [78]. While microbial H_2 generation is less extensive compared to abiotic processes such as serpentinization, it contributes to sustaining H_2 concentrations in organic-rich geological systems.

The genesis of natural hydrogen in subsurface gas reservoirs is commonly attributed to biological processes. Documented microbial pathways include anaerobic organic matter degradation, fermentative metabolism, and nitrogenase-mediated production by diazotrophic bacteria [93–95]. Controlled laboratory studies have reproducibly demonstrated H_2 generation by diverse microbial taxa under substrate-rich conditions, as extensively cataloged in prior syntheses [96]. Despite robust experimental evidence for microbial hydrogen genesis, field observations universally indicate that H_2-producing microorganisms exist in obligate syntrophic associations with hydrogenotrophic archaea and bacteria [97]. This interspecies metabolic coupling drives immediate scavenging of biologically generated H_2 through methanogenesis, acetogenesis, or sulfate reduction [93, 98]. The evolutionary persistence of such consortia reflects the thermodynamic constraints imposed by H_2 accumulation on anaerobic respiration energetics [99]. Contemporary geomicrobiological frameworks, as systematically synthesized in recent reviews [98], explicitly integrate these ecological dynamics with geochemical patterns of H_2 distribution in crustal systems.

The critical focus in this field should shift from H_2 generation to its utilization, a perspective that may elucidate the observed scarcity of H_2 in ideal reservoirs via biological exclusion mechanisms. Two primary decomposition pathways have been recognized: microbial metabolism and soil-mediated reactions [100]. Soil-driven H_2 uptake is responsible for approximately 80% of atmospheric H_2 removal [101, 102]. The surface soil horizon represents the primary zone for H_2 turnover, coinciding with the highest concentration of microbial biomass in terrestrial environments [94]. Notably, soil systems exhibit no detectable endogenous H_2 production, with measured emission rates being orders of magnitude lower than simultaneous consumption rates [103]. Prevailing theories posit that nonbiological mechanisms may supersede microbial processes in governing H_2 decomposition dynamics [100].

Research across diverse soil types demonstrates that H_2 production is particularly substantial in wetland soils, though most of the produced H_2 is quickly metabolized into CH_4 by methanogenic archaea [94]. In contrast, aerobic upland soils primarily generate H_2 through nitrogen-fixing bacteria, with consumption driven by non-biological enzymatic pathways rather than microbial activity. Conversely, in anaerobic wetland soils, bacterial fermentation serves as the principal mechanism for hydrogen production, which is subsequently consumed by methanogens, sulfate reducers, denitrifiers, iron-reducing bacteria, and Knall gas bacteria [94, 104]. Studies indicate that more than 90% of H_2 is depleted within 1 cm of its origin, and its accumulation significantly impairs the soil's oxidative capacity for reduced gases such as CH_4 and CO [105].

The long-standing assumption that atmospheric H_2 concentrations are too low to support bacterial growth, despite enzymatic activity rapidly scavenging trace H_2 from air, has underpinned decades of soil uptake research [94]. However, subsequent studies by Constant et al. challenged this paradigm by isolating an aerobic bacterium (*Streptomyces* sp.) capable of oxidizing H_2 at ambient tropospheric levels, suggesting that metabolic activity in living cells, rather than extracellular enzymes, drives soil H_2 uptake [101]. Their work further revealed widespread H_2 consumption activity at atmospheric concentrations across Streptomyces species, which thrive in diverse ecosystems such as deserts, forests, and peatlands. These findings suggest that specific H_2-oxidizing actinobacteria serve as the principal biological sink within the global atmospheric hydrogen cycle [101]. Recent studies have revealed that specific microbial communities in Antarctic environments rely solely on atmospheric hydrogen as their primary energy source [106].

Hence, soils act as the principal reservoir for atmospheric H_2 budgets. Previous research suggests that soil moisture predominantly governs H_2 consumption rates, with minimal temperature dependence [103]. Contemporary investigations reveal that moderately humid soils achieve peak H_2 uptake near 30 °C [107]. Remarkably, soils continue absorbing H_2 below freezing, as hydrogenase enzymes remain catalytically active under such conditions. This finding provides robust evidence for microbially-driven H_2 consumption through bacterial metabolic processes [107].

Soil H_2 consumption capacity increases significantly with rising H_2 concentrations. When concentrations approach the lower explosive limit (LEL, ~4%), soil consumption rates exhibit a 1000-fold enhancement [108]. This results from the activity of two physiologically distinct communities of H_2-oxidizing microbes that exhibit differential kinetics of H_2 consumption [101]. Experimental studies indicate that predictive models of soil H_2 uptake capacity should integrate total carbon content with the relative abundance of H_2-oxidizing bacteria as primary variables [109]. Additionally, H_2 oxidation processes have been observed to take place within rhizosphere environments and root systems [94]. H_2 uptake intensity, quantified as deposition velocity, varies between 0.06 to 0.22 cm s^{-1} across different soil types [110]. Comprehensive compilations of H_2 deposition velocities have been reported for diverse ecosystems [111]. Published estimates of global soil H_2 consumption rates include: 120 Tg yr^{-1} [112], 56 ± 41 Tg yr^{-1} [113], 90 ± 20 Tg yr^{-1} [95, 100, 104], 88 ± 11 Tg yr^{-1} [102], and $60 + 30 - 20$ Tg yr^{-1} [111].

Research in the scientific literature has provided estimates of biological H_2 production. As soil acts as the main H_2 reservoir, differentiating atmospheric emissions from microbial and enzymatic consumption within the soil remains a challenge. Earlier research indicates that soil microbial H_2 production contributes minimally to the atmospheric H_2 budget [94]. Field studies on H_2 emissions from nitrogen-fixing legumes have led to estimates of 2.4–4.9 Tg/yr for biological nitrogen fixation [95, 114]. This value considerably surpasses previous estimates of global biological hydrogen production, which amounted to 0.017 Tg/yr from sources including paddy fields, enteric fermentation in animals, and other natural ecosystems [115]. Recent investigations reveal that roughly 97% of H_2 produced by nitrogen-fixing bacteria undergoes removal via microbial consumption and soil chemistry before reaching the atmosphere [110].

Conversely, organic fermentation is incapable of achieving high H_2 yields. The viability of producing H_2 via the fermentation of complex organic substrates was assessed through the calculation of Gibbs free energy. However, the analysis concluded that microbial fermentation is not a viable mechanism to account for the elevated H_2 concentrations detected [2].

Interestingly, H_2 production has also been documented in termites and some protozoa. Nevertheless, no studies have been published regarding metabolic H_2 production in higher organisms [94]. Estimates suggest that termites may release up to 200 Tg of H_2 per year [116], a substantial volume compared to other natural hydrogen sources, although this figure is contested by some studies as potentially overestimated [111]. A significant portion of this hydrogen is presumably rapidly assimilated in situ through symbiotic relationships with methanogenic archaea. It has been proposed that only a fraction of the produced H_2, approximately 1 Tg/yr, might be released into the atmosphere [111].

It has been frequently reported that H_2 concentrations in seawater exhibit supersaturation relative to atmospheric equilibrium (i.e., dissolved H_2 concentrations exceeding atmospheric equilibrium values) [117]. H_2 concentrations as high as 7.3 cm^3/L have been documented in Caspian Sea waters [118], though the source of this gas remains unexplained. Certain Atlantic water masses demonstrate threefold supersaturation [119]. These observations collectively suggest that marine environments should constitute a source of atmospheric H_2. Current estimates of oceanic H_2 production range from 3 ± 2 Tg/yr to 4 ± 2 Tg/yr and 6 ± 3 Tg/yr [104, 111]. Recent isotopic analyses of dissolved marine H_2 propose production mechanisms involving nitrogen-fixing organisms under "significantly distinct source inputs" [120]. The authors hypothesize potential photochemical origins while acknowledging possible geological contributions. Substantiating this, another study observed no diurnal variation in H_2 concentrations, thereby challenging photochemical and biological source hypotheses [117].

Research has shown cases where H_2-producing microorganisms were not present in samples collected from H_2-rich wells, as documented in this reference [121]. A separate study investigated the distribution and activity of H_2-producing and H_2-consuming bacteria in deep water samples, revealing that only minimal microbial activity was detected across 13 probes with dissolved H_2 [71]. Furthermore, no bacterial activity was identified in H_2-enriched groundwater near petroleum reservoirs [122]. By contrast, only one investigation has indicated that the H_2 detected in a restricted set of samples originated from microbial activity within drilling mud [123].

It is important to recognize that in deep well environments, elevated temperatures and pressures often exceed the limits for microbial survival; therefore, any hydrogen detected in samples from such settings is unlikely to originate from bacterial processes [121]. However, there are increasing examples of H_2-based deep microbial communities inhabiting oceanic depths and crustal fractures. The H_2 consumption rates of these organisms remain unquantified, though they likely exceed diffusion rates. Under such circumstances, the low H_2 levels detected across numerous geological settings may be attributed to microbial consumption processes [124] (Table 4.1).

Table 4.1 H_2 generation and consumption under the mechanism of microbial activity

Process	Reaction	References
Microbial hydrogen generation		
Fermentation	Multiple pathways that break down large organics into smaller organics, for example, mixed acid fermentation, e.g., $C_6H_{12}O_6 + 4H_2O \rightarrow 2CH_3COO_3^- + 2HCO_3^- + 4H^+ + 4H_2$	[98, 125–127]
Nitrogen fixation (nitrogenase activity)	$N_2 + 8H^+ + 8e^-(Fd_{red}) \rightarrow 2NH_3 + H_2(+Fd_{ox})$	[98, 125–127]
Anaerobic carbon monoxide oxidation	$CO + H_2O \rightarrow CO_2 + H_2$	[98, 125–127]
Phosphite oxidation	$H_3PO_3 + H_2O \rightarrow H_3PO_4 + H_2$	[98, 125–127]
Acetate oxidation	$\frac{1}{4}CH_3COO^- + 14H^+ + \frac{1}{2}H_2O \rightarrow H_2 + \frac{1}{2}CO_2$	[98, 125–127]
Microbial hydrogen consumption		
Hydrogenotrophic methanogenesis	$HCO3^- + H_2 + \frac{1}{4}H+ \rightarrow \frac{1}{4}CH_4 + \frac{3}{4}H_2O$	[98, 127–129]
Acetogenesis	$\frac{1}{2}HCO3^- + H_2 + \frac{1}{4}H^+ \rightarrow \frac{1}{4}\ CH_3COO - +2H_2O$	[98, 127–129]
Sulfate reduction	$\frac{1}{4}SO_4^{2-} + H_2 + \frac{1}{4}H^+ \rightarrow \frac{1}{4}HS - +H_2O$	[98, 127–129]
Sulfur reduction	$H_2 + S \rightarrow H_2S$	[98, 127–129]
Iron (III) reduction	$2FeOOH + H_2 + 4H^+ \rightarrow 2Fe^{2+} + 4H_2O$	[98, 127–129]
Aerobic hydrogen oxidation	$H_2 + \frac{1}{2}O_2 \rightarrow H_2O$	[98, 127–129]
Dehalorespiration	Halogenated compounds + $H_2 \rightarrow$ dehalogenated compounds + 2HCL	[98, 127–129]
Fumarate respiration	H_2 + fumarate $\rightarrow$ succinate	[98, 127–129]
Denitrification	$\frac{2}{5}NO_3^{-1} + H_2 + \frac{2}{5}H^+ \rightarrow \frac{1}{5}N_2 + \frac{6}{5}H_2O$	[98, 127–129]

References

1. Zgonnik, V. 2020. The occurrence and geoscience of natural hydrogen: A comprehensive review. *Earth-Science Reviews* 203: 103140.
2. Lin, L.H., et al. 2005. Radiolytic H_2 in continental crust: Nuclear power for deep subsurface microbial communities. *Geochemistry Geophysics Geosystems* 6.

3. Osselin, F., et al. 2022. Orange hydrogen is the new green. *Nature Geoscience* 15 (10): 765–769.
4. Dou, L., et al. 2024. Global natural hydrogen exploration and development situation and prospects in China. *Lithologic Reservoirs* 36 (02): 1–14 (in Chinese).
5. Prinzhofer, A., C. S. T. Cissé, and A. B. Diallo. 2018. Discovery of a large accumulation of natural hydrogen in Bourakebougou (Mali). *International Journal of Hydrogen Energy* 43 (42): 19315–19326.
6. Bachaud, P., et al. 2017. Modeling of hydrogen genesis in ophiolite massif. *Procedia Earth and Planetary Science* 17: 265–268.
7. Suo, Y., et al. 2024. Oean-floor hydrogen accumulation model and global distribution. *Earth Science Frontiers* 31 (04): 175–182 (in Chinese).
8. Fang, D., et al. 2024. Genesis of natural hydrogen and progress in global hydrogen resource exploration. *Journal of Chengdu University of Technology(Science & Technology Edition)*, 1–24 (in Chinese).
9. Liu, Q.Y., et al. 2024. Natural hydrogen: A potential carbon-free energy source. *Chinese Science Bulletin-Chinese* 69 (17): 2344–2350 (in Chinese).
10. Hand, E. 2023. Hidden hydrogen: Does Earth hold vast stores of a renewable, carbon-free fuel? *Science* 379: 630–636.
11. Milkov, A. V. 2022. Molecular hydrogen in surface and subsurface natural gases: Abundance, origins and ideas for deliberate exploration. *Earth-Science Reviews* 230: 104063.
12. Giuliani, A., et al. 2023. Genesis and evolution of kimberlites. *Nature Reviews Earth & Environment* 4 (11): 738–753.
13. Allen, D. E., and W. E. Seyfried. 2004. Serpentinization and heat generation: Constraints from Lost City and Rainbow hydrothermal systems. *Geochimica et Cosmochimica Acta* 68 (6): 1347–1354.
14. Malachowska, A., et al. 2022. Hydrogen storage in geological formations—The potential of salt caverns. *Energies* 15 (14).
15. Wei, Q., et al. 2023. Geological characteristics, formation distribution and resource prospects of natural hydrogen reservoir. *Natural Gas Geoscience* 35 (6): 1113–1122 (in Chinese).
16. Jackson, O., et al. 2024. Natural hydrogen: Sources, systems and exploration plays. *Geoenergy* 2 (1).
17. Meng, Q., et al. 2015. Distribution and geochemical characteristics of hydrogen in natural gas from the Jiyang Depression, Eastern China. *Acta Geologica Sinica (English Edition)* 89 (5): 1616–1624.
18. Wang, L., et al. 2023. The origin and occurrence of natural hydrogen. *Energies* 16 (5): 2400–2400.
19. Tian, H., et al. 2019. Organic waste to biohydrogen: A critical review from technological development and environmental impact analysis perspective. *Applied Energy* 256: 113961.
20. Maiga, O., et al. 2023. Characterization of the spontaneously recharging natural hydrogen reservoirs of Bourakebougou in Mali. *Scientific reports* 13 (1): 11876–11876.
21. Klein, F., J. D. Tarnas, and W. Bach. 2020. Abiotic sources of molecular hydrogen on Earth. *Elements* 16 (1): 19–24.
22. Blay-Roger, R., et al. 2024. Natural hydrogen in the energy transition: Fundamentals, promise, and enigmas. *Renewable and Sustainable Energy Reviews* 189: 113888.
23. Tian, Q., et al. 2022. Origin, discovery, exploration and development status and prospect of global natural hydrogen under the background of "carbon neutrality." *China Geology* 5 (4): 722–733.
24. Worman, S.L. 2015. *Global Rates of Free Hydrogen (H_2) Production by Serpentinization and Other Abiogenic Processes Within Young Ocean Crust*. Duke University.

25. Holloway, J.R., and P.A. O'DAY. 2000. Production of CO_2 and H_2 by diking-eruptive events at mid-ocean ridges: Implications for abiotic organic synthesis and global geochemical cycling. *International Geology Review* 42 (8): 673–683.
26. Lollar, B. S., et al. 2014. The contribution of the Precambrian continental lithosphere to global H_2 production. *Nature* 516 (7531): 379–382.
27. Mayhew, L. E., et al. 2013. Hydrogen generation from low-temperature water-rock reactions. *Nature Geoscience* 6 (6): 478–484.
28. Ross, C.M., et al. 2025. Hydrogen generation and serpentinization of olivine under flow conditions. *Geophysical Research Letters* 52 (6).
29. Tian, Q., et al. 2022. Non-negligible new energy in the energy transition context: Natural hydrogen. *Geological Survey of China* 9 (01): 1–15 (in Chinese).
30. Boreham, C.J., et al. 2021. Hydrogen in Australian natural gas: Occurrences, sources and resources. *The Australian Energy Producers Journal* 61 (1): 163–191.
31. McCollom, T. M., and J. S. Seewald. 2013. Serpentinites, hydrogen, and life. *Elements* 9 (2): 129–134.
32. Worman, S. L., et al. 2016. Global rate and distribution of H_2 gas produced by serpentinization within oceanic lithosphere. *Geophysical Research Letters* 43 (12): 6435–6443.
33. Brovarone, A.V., et al. 2020. Subduction hides high-pressure sources of energy that may feed the deep subsurface biosphere. *Nature Communications* 11 (1).
34. Şimşek, E., O. Parlak, and A. H. F. Robertson. 2023. Ion-probe (SIMS) U-Pb geochronology and geochemistry of the Upper Cretaceous Kızıldağ (Hatay) ophiolite: Implications for supra-subduction zone spreading in the Southern Neotethys. *Geosystems and Geoenvironment* 2 (3): 100165.
35. Morag, N., et al. 2020. The origin of plagiogranites: Coupled SIMS O isotope ratios, U–Pb dating and trace element composition of zircon from the Troodos ophiolite, Cyprus. *Journal of Petrology* 61 (5): egaa057.
36. Ulrich, M., et al. 2020. Serpentinization of New Caledonia peridotites: From depth to (sub-) surface. *Contributions to Mineralogy and Petrology* 175 (9): 208–220.
37. McCollom, T. M., and W. Bach. 2009. Thermodynamic constraints on hydrogen generation during serpentinization of ultramafic rocks. *Geochimica et Cosmochimica Acta* 73 (3): 856–875.
38. McCollom, T. M., and C. Donaldson. 2016. Generation of hydrogen and methane during experimental low-temperature reaction of ultramafic rocks with water. *Astrobiology* 16 (6): 389–406.
39. Sleep, N. H., et al. 2004. H_2-rich fluids from serpentinization: Geochemical and biotic implications. *Proceedings of the National Academy of Sciences of the United States of America* 101 (35): 12818–12823.
40. Rezaee, R. 2025. Quantifying natural hydrogen generation rates and volumetric potential in onshore serpentinization. *Geosciences* 15 (3): 112.
41. Canfield, D. E., M. T. Rosing, and C. Bjerrum. 2006. Early anaerobic metabolisms. *Philosophical Transactions of the Royal Society B: Biological Sciences* 361 (1474): 1819–1836.
42. Sleep, N. H., and D. K. Bird. 2007. Niches of the pre-photosynthetic biosphere and geologic preservation of Earth's earliest ecology. *Geobiology* 5 (2): 101–117.
43. Bach, W., and K. J. Edwards. 2003. Iron and sulfide oxidation within the basaltic ocean crust: Implications for chemolithoautotrophic microbial biomass production. *Geochimica et Cosmochimica Acta* 67 (20): 3871–3887.
44. Zgonnik, V., et al. 2019. Diffused flow of molecular hydrogen through the Western Hajar mountains, Northern Oman. *Arabian Journal of Geosciences* 12 (3): 71.

45. Charlou, J. L., et al. 2002. Geochemistry of high H_2 and CH_4 vent fluids issuing from ultramafic rocks at the Rainbow hydrothermal field (36° 14′ N, MAR). *Chemical Geology* 191 (4): 345–359.
46. Cannat, M., F. Fontaine, and J. Escartin. 2010. Serpentinization and hydrothermal venting at slow-spreading ridges. *Geochimica Et Cosmochimica Acta* 74 (12): A138–A138.
47. Keir, R.S. 2010. A note on the fluxes of abiogenic methane and hydrogen from mid-ocean ridges. *Geophysical Research Letters* 37.
48. Sherwood Lollar, B. 2004. Geochemistry. Life's chemical kitchen. *Science (New York, N.Y.)* 304 (5673): 972–973.
49. Stevens, T. O., and J. P. McKinley. 2000. Abiotic controls on H_2 production from basalt-water reactions and implications for aquifer biogeochemistry. *Environmental Science & Technology* 34 (5): 826–831.
50. Anderson, D, 1998. *Studies of the Earth's Deep Interior*. National Science Foundation.
51. Milesi, V., et al. 2016. Contribution of siderite-water interaction for the unconventional generation of hydrocarbon gases in the Solimoes basin, north-west Brazil. *Marine and Petroleum Geology* 71: 168–182.
52. Meng, Q., et al. 2021. Geological background and exploration prospects for the occurrence of high-content hydrogen. *Petroleum Geology & Experiment* 43 (02): 208–216 (in Chinese).
53. Song, H. 2022. *Mechanism of $Fe(OH)_2$ Phase Transformation to Produce H_2 During Low Temperature Serpentinization*. South China University of Technology (in Chinese).
54. Tarapoanca, M., et al. 2013. Forward kinematic modelling of a regional transect in the Northern Emirates using geological and apatite fission track age constraints on paleo-burial history. In *Lithosphere dynamics and sedimentary basins: The Arabian plate and analogues*. Springer.
55. Sallarès, V., et al. 2013. Overriding plate structure of the Nicaragua convergent margin: Relationship to the seismogenic zone of the 1992 tsunami earthquake. *Geochemistry, Geophysics, Geosystems* 14 (9): 3436–3461.
56. Hutchinson, I.P., et al. 2024. Greenstones as a source of hydrogen in cratonic sedimentary basins. *Geological Society, London, Special Publications* 547 (1): SP547-2023-39.
57. Vacquand, C., et al. 2018. Reduced gas seepages in ophiolitic complexes: Evidences for multiple origins of the H_2-CH_4-N_2 gas mixtures. *Geochimica et Cosmochimica Acta* 223: 437–461.
58. Bourdet, J., et al. 2023. Natural hydrogen in low temperature geofluids in a Precambrian granite, South Australia. Implications for hydrogen generation and movement in the upper crust. *Chemical Geology*. 638.
59. Sartorio, C., et al. 2022. Preliminary assessment of radiolysis for the cooling water system in the rotating target of SORGENTINA-RF. *Environments* 9: 106.
60. Dubessy, J., et al. 1988. Radiolysis evidenced by H_2-O_2 and H_2-bearing fluid inclusions in three uranium deposits. *Geochimica et Cosmochimica Acta* 52 (5): 1155–1167.
61. Вовк, И.Ф. 1979. *Радиолиз подземных вод и его геохимическая роль*. Nedra.
62. Blair, C. C., et al. 2007. Radiolytic hydrogen and microbial respiration in subsurface sediments. *Astrobiology* 7 (6): 951–970.
63. Vovk, I.F. 1982. Radiolysis of underground waters as the mechanism of geochemical transformation of the energy of radioactive decay in sedimentary rocks. *Lithology and Mineral Resources* 16.
64. Parnell, J., and N. Blamey. 2017. Hydrogen from radiolysis of aqueous fluid inclusions during diagenesis. *Minerals* 7(8).
65. McMahon, S., J. Parnell, and N. J. F. Blamey. 2016. Evidence for seismogenic hydrogen gas, a potential microbial energy source on Earth and Mars. *Astrobiology* 16 (9): 690–702.
66. Kubatko, K. A. H., et al. 2003. Stability of peroxide-containing uranyl minerals. *Science* 302 (5648): 1191–1193.

67. Savchenko, V. P. 1958. The formation of free hydrogen in the Earth's crust as determined by the reducing action of the products of radioactive transformations of isotopes. *Geochemistry International* 1: 16–25.
68. Вовк, ИФ. 1978. О природе воды в месторождениях калийных солей. *Geokhimiya* 1: 122–127.
69. Морачевский, Ю, А Самарцева, and А Черепенников. 1937. Газоносность толщи калийных солей Верхнекамского месторождения. *Калий* 7: 24–31.
70. Nesmelova, Z. N., and L. G. Travnikova. 1973. Radiogenic gases in ancient salt deposits. *Geochemistry International* 10: 554–559.
71. Стадник, Е. 1970. О содержании водорода в газах, растворенных в подземных водахсеверо-западного обращения. *Актуальные проблемы нефти и газа* 1: 36–44.
72. He, H., et al. 2021. An abiotic source of Archean hydrogen peroxide and oxygen that predates oxygenic photosynthesis. *Nature Communications* 12 (1): 6611.
73. Nivin, V. A. 2009. Diffusively disseminated hydrogen-hydrocarbon gases in rocks of nepheline syenite complexes. *Geochemistry International* 47 (7): 672–691.
74. Deloule, E., F. Albarede, and S. M. Sheppard. 1991. Hydrogen isotope heterogeneities in the mantle from ion probe analysis of amphiboles from ultramafic rocks. *Earth and Planetary Science Letters* 105 (4): 543–553.
75. Schmandt, B., et al. 2014. Dehydration melting at the top of the lower mantle. *Science* 344 (6189): 1265–1268.
76. Smith, E. M., et al. 2016. Large gem diamonds from metallic liquid in Earth's deep mantle. *Science* 354 (6318): 1403–1405.
77. Briere, D., T. Jerzykiewicz, and W. Śliwiński. 2016. On generating a geological model for hydrogen gas in the Southern Taoudenni Megabasin. (Bourakebougou Area, Mali).
78. Boileau, C., et al. 2016. Hydrogen production by the hyperthermophilic bacterium Thermotoga maritima part I: Effects of sulfured nutriments, with thiosulfate as model, on hydrogen production and growth. *Biotechnology for Biofuels* 9 (1): 269.
79. Sandatlas. 2011. *Limestone—Sedimentary Rocks.* Sandatlas.org.
80. Ikorsky, S.V., et al. 1999. The investigation of gases during the Kola Superdeep Borehole drilling (to 11.6 km depth). (107), 145–152.
81. Shcherbakov, A. V., and N. D. Kozlova. 1986. Occurrence of hydrogen in subsurface fluids and the relationship of anomalous concentrations to deep faults in the USSR. *Geotectonics* 20: 120–128.
82. Yang, X., H. Keppler, and Y. Li. 2016. Molecular hydrogen in mantle minerals. *Geochemical Perspectives Letters* 2: 160–168.
83. McCarthy Jr. J.H., et al. 1986. Soil gas studies around hydrogen-rich natural gas wells in northern Kansas. In *Open-File Report.*
84. Rohrbach, A., et al. 2011. Experimental evidence for a reduced metal-saturated upper mantle. *Journal of Petrology* 52 (4): 717–731.
85. Murphy, C.A. 2016. Hydrogen in the Earth's core. In *Deep Earth*, 253–264.
86. Poirier, J.-P. 1994. Light elements in the Earth's outer core: A critical review. *Physics of the Earth and Planetary Interiors* 85 (3): 319–337.
87. Tagawa, S., et al. 2016. Compression of Fe-Si-H alloys to core pressures. *Geophysical Research Letters* 43 (8): 3686–3692.
88. Larin, V.N. 1993. *Hydridic Earth: The new geology of our primordially hydrogen-rich planet.*
89. Mukhopadhyay, S. 2012. Early differentiation and volatile accretion recorded in deep-mantle neon and xenon. *Nature* 486 (7401): 101-U124.
90. Serovaiskii, A., and V. Kutcherov. 2020. Formation of complex hydrocarbon systems from methane at the upper mantle thermobaric conditions. *Scientific Reports* 10 (1): 4559.

91. Bardin, D. J. 1987. Deep gas theory as an exploratory tool. *Energy Exploration & Exploitation* 5: 255–264.
92. Boruah, A., P. Alaktaraag, and S. Singh. 2024. Utilization of coal for hydrogen generation. *International Journal of Coal Preparation and Utilization*, 1–22.
93. Morita, R. Y. 1999. Is H_2 the universal energy source for long-term survival? *Microbial Ecology* 38 (4): 307–320.
94. Conrad, R. 1996. Soil microorganisms as controllers of atmospheric trace gases (H_2, CO, CH_4, OCS, N_2O, and NO). *Microbiological Reviews* 60 (4): 609–640.
95. Conrad, R., and W. Seiler. 1980. Contribution of hydrogen production by biological nitrogen fixation to the global hydrogen budget. *Journal of Geophysical Research: Oceans* 85 (C10): 5493–5498.
96. Nandi, R., and S. Sengupta. 1998. Microbial production of hydrogen: An overview. *Critical Reviews in Microbiology* 24 (1): 61–84.
97. Nealson, K. H., F. Inagaki, and K. Takai. 2005. Hydrogen-driven subsurface lithoautotrophic microbial ecosystems (SLiMEs): Do they exist and why should we care? *Trends in Microbiology* 13 (9): 405–410.
98. Gregory, S.P., et al. 2019. Subsurface microbial hydrogen cycling: Natural occurrence and implications for industry. *Microorganisms* 7 (2).
99. Hoehler, T.M., 2005. Biogeochemistry of dihydrogen (H_2). In *Biogeochemical Cycles of Elements*, ed. A. Sigel, H. Sigel, and R.K.O. Sigel, 9–48.
100. Conrad, R., and W. Seiler. 1981. Decomposition of atmospheric hydrogen by soil microorganisms and soil enzymes. *Soil Biology and Biochemistry* 13 (1): 43–49.
101. Constant, P., et al. 2010. Streptomycetes contributing to atmospheric molecular hydrogen soil uptake are widespread and encode a putative high-affinity NiFe-hydrogenase. *Environmental Microbiology* 12 (3): 821–829.
102. Rhee, T. S., C. A. M. Brenninkmeijer, and T. Röckmann. 2006. The overwhelming role of soils in the global atmospheric hydrogen cycle. *Atmospheric Chemistry and Physics* 6: 1611–1625.
103. Conrad, R., and W. Seiler. 1985. Influence of temperature, moisture, and organic carbon on the flux of H_2 and CO between soil and atmosphere: Field studies in subtropical regions. *Journal of Geophysical Research: Atmospheres* 90 (D3): 5699–5709.
104. Seiler, W., and R. Conrad. 1987. Contribution of tropical ecosystems to the global budgets of trace gases, especially CH_4, H_2, CO and N_2O. In *The Geophysiology of Amazonia Vegetation and Climate Interactions*, 133–162.
105. Piché-Choquette, S., M. Khdhiri, and P. Constant. 2018. Dose-response relationships between environmentally-relevant H_2 concentrations and the biological sinks of H_2, CH_4 and CO in soil. *Soil Biology & Biochemistry* 123: 190–199.
106. Ji, M., et al. 2017. Atmospheric trace gases support primary production in Antarctic desert surface soil. *Nature* 552 (7685): 400-+.
107. Ehhalt, D. H., and F. Rohrer. 2011. The dependence of soil H_2 uptake on temperature and moisture: A reanalysis of laboratory data. *Tellus Series B-Chemical and Physical Meteorology* 63 (5): 1040–1051.
108. Sukhanova, N. I., et al. 2013. Changes in the humus status and the structure of the microbial biomass in hydrogen exhalation places. *Eurasian Soil Science* 46 (2): 135–144.
109. Khdhiri, M., et al. 2015. Soil carbon content and relative abundance of high affinity H_2-oxidizing bacteria predict atmospheric H_2 soil uptake activity better than soil microbial community composition. *Soil Biology & Biochemistry* 85: 1–9.
110. Chen, Q., et al. 2015. Isotopic signatures of production and uptake of H_2 by soil. *Atmospheric Chemistry and Physics* 15 (22): 13003–13021.

111. Ehhalt, D. H., and F. Rohrer. 2009. The tropospheric cycle of H_2: A critical review. *Tellus Series B-Chemical and Physical Meteorology* 61 (3): 500–535.
112. Herbert, E.H. 1981. *Geothermal Hydrogen—1980 Jackling Lecture.* The American Institute of Mining, Metallurgical, and Petroleum Engineers.
113. Novelli, P., et al. 1999. Molecular hydrogen in the troposphere: Global distribution and budget. *Journal of Geophysical Research* 104: 30427–30444.
114. Conrad, R., and W. Seiler. 1979. Field measurements of hydrogen evolution by nitrogen-fixing legumes. *Soil Biology and Biochemistry.* 11 (6): 689–690.
115. Koyama, T. 1963. Gaseous metabolism in lake sediments and paddy soils and the production of atmospheric methane and hydrogen. *Journal of Geophysical Research (1896–1977)* 68 (13): 3971–3973.
116. Zimmerman, P. R., et al. 1982. Termites: A potentially large source of atmospheric methane, carbon dioxide, and molecular hydrogen. *Science* 218 (4572): 563–565.
117. Bullister, J. L., N. L. Guinasso Jr., and D. R. Schink. 1982. Dissolved hydrogen, carbon monoxide, and methane at the CEPEX site. *Journal of Geophysical Research: Oceans* 87 (C3): 2022–2034.
118. Levshounova, S. P. 1991. Hydrogen in petroleum geochemistry. *Terra Nova* 3 (6): 579–585.
119. Schmidt, U. 1974. Molecular hydrogen in the atmosphere. *Tellus* 26 (1–2): 78–90.
120. Walter, S., et al. 2016. Isotopic evidence for biogenic molecular hydrogen production in the Atlantic Ocean. *Biogeosciences* 13 (1): 323–340.
121. Войтов, Г., and Д. Осика. 1982. Водородное дыхание Земли как отражение особеннойгеологического строения и театонического развития ее мегаструктур. Альтернативная энергетика и экология, pp. 7–29.
122. Зингер, А. 1962. Молекулярный водород в составе газа, растворенного в водах газонефтяных месторождений Нижнего Поволжья. *Геохимия* 10: 890–898.
123. Несмелова, З, and Е Рогозина. 1963. К вопросу о водороде в природных газах. *Уголь* 4: 27–35.
124. Fedonkin, M. A. 2009. Eukaryotization of the early biosphere: A biogeochemical aspect. *Geochemistry International* 47 (13): 1265–1333.
125. Conrad, R. 1999. Contribution of hydrogen to methane production and control of hydrogen concentrations in methanogenic soils and sediments. *FEMS Microbiology Ecology* 28 (3): 193–202.
126. Thauer, R. K., K. Jungermann, and K. Decker. 1977. Energy conservation in chemotrophic anaerobic bacteria. *Bacteriological Reviews* 41 (1): 100–180.
127. Schink, B., et al. 2002. Desulfotignum phosphitoxidans sp. nov., a new marine sulfate reducer that oxidizes phosphite to phosphate. *Archives of Microbiology* 177 (5): 381–391.
128. Greening, C., et al. 2016. Genomic and metagenomic surveys of hydrogenase distribution indicate H_2 is a widely utilised energy source for microbial growth and survival. *The ISME Journal* 10 (3): 761–777.
129. Dopffel, N., S. Jansen, and J. Gerritse. 2021. Microbial side effects of underground hydrogen storage—Knowledge gaps, risks and opportunities for successful implementation. *International Journal of Hydrogen Energy* 46 (12): 8594–8606.

Migration and Accumulation 5

5.1 Migration

H_2 migration primarily occurs along faults and fractures within the Earth's crust, which act as conduits for its upward transport from deep-seated sources (Fig. 5.1). However, different from hydrocarbon migration, which is typically characterized by a large-scale transport and accumulation, natural hydrogen migration occurs as a rapid-diffusion, dynamic-equilibrium process over short timescales. This fundamental distinction accounts for differences in resource accumulation scale and longevity between the two systems, and also explains why natural hydrogen occurrences are often small-scale and discretely-distributed. Regarding the driving force of the molecules, migration of hydrogen molecules is influenced by driving forces such as buoyancy-driven, pressure gradients, diffusion-driven and convection-driven, which are similar as that of hydrocarbons [1, 2]. Among them, diffusion-driven transport represents the most principal migration mechanism for natural hydrogen. Within the matrix pore structures, hydrogen molecules can rapidly overcome capillary pressure barriers, enabling homogeneous micro-scale percolation. This is due to the extremely small hydrogen molecular radius and diffusion coefficients, which are over three times higher than those of hydrocarbons [3, 4]. The migration pathways of natural hydrogen are primarily governed by the combined influence of geological structures and lithological conditions. Research has demonstrated that H_2 concentrations are frequently associated with fault systems, manifesting in distinct geological structures such as the "fairy circles" documented globally. A notable example is the Darling Fault in northern Western Australia, where H_2 accumulations exemplify the direct link between deep fault zones and H_2 occurrence. Hydrothermal systems also play a critical role in H_2 migration, as elevated geothermal gradients promote the circulation of H_2-enriched fluids. Furthermore, tectonic processes interacting with local geology govern the pathways and efficiency of H_2 migration [5, 6]. In the Mali region, H_2 migration

Y. Yuan et al., *Natural Hydrogen*, Synthesis Lectures on Renewable Energy Technologies, https://doi.org/10.1007/978-3-032-14469-0_5

dynamics are influenced by fracture density, porosity variations, and the structural architecture of dolerite formations, facilitating its ascent from deep-seated origins to shallower reservoirs (Fig. 5.2) [7].

Deep-seated fluids can transport substantial amounts of H_2 to shallow crustal levels, with distinct variations in H_2 transport capacities among different fluid types. Generally, 1 kg of rock from 15 km depth can release approximately 75 cm^3 of H_2 when reaching the surface. Specifically, alkaline rocks exhibit an average H_2 content of 3 cm^3/kg, whereas mafic-ultramafic rocks demonstrate significantly higher H_2 transport capacity with 26.8 cm^3/kg [8]. Notably, 1 kg of water migrating from 15 km depth to the surface can release up to 12,000 cm^3 of H_2, indicating that aqueous fluids serve as highly efficient H_2 carriers. Consequently, deep-seated fluid activity zones, particularly those associated with mafic and ultramafic volcanic terrains, represent prime areas for H_2 enrichment.

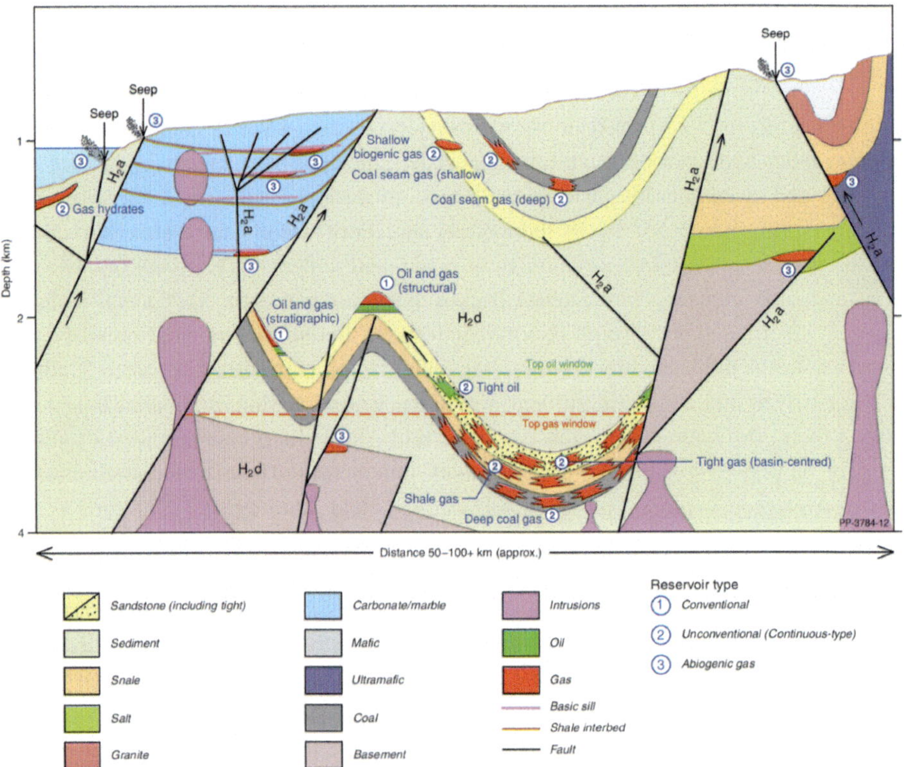

Fig. 5.1 Geological model of source-migration-accumulation for natural hydrogen resource and migration in both conventional and non-conventional systems. Hydrogen concentrations in petroleum systems (1, 2) are likely enriched by contributions from abiogenic hydrogen sources (3). Note that H_2a denotes the advective migration of hydrogen, whereas H_2d refers to the diffusive migration of hydrogen (reproduced with permission from [17])

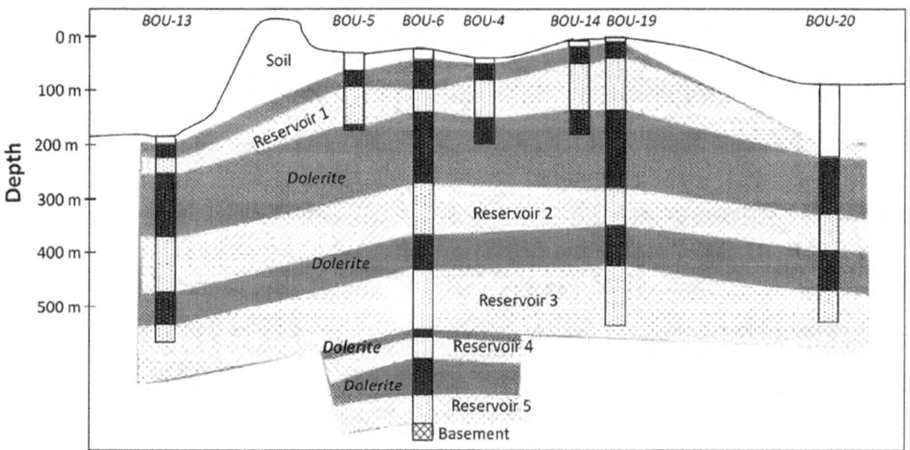

Fig. 5.2 Conceptual geological model of the Bourakebougou H_2-bearing field (Mali), showing interpreted intervals of potential hydrogen accumulation and migration ([12])

The formation of these H_2-rich reservoirs is fundamentally constrained by regional tectonic settings, as structural frameworks govern both fluid migration pathways and H_2 accumulation mechanisms [9].

Microstructural elements (e.g., lamination surfaces, cleavage foliations, and lineation trajectories [10, 11]) exert a crucial control on the migration pathway of molecular H_2, particularly in settings characterized by vertically -oriented structural discontinuities. In lithostratigraphic units such as the Great Cobar Slate and CSA Siltstone, these fabrics define steeply dipping conduits, interconnected pathways that facilitate the upward migration of hydrogen. Discordant intrusions, like dikes, generate chemically and thermally altered contact zones, enabling H_2 migration through these structurally weakened pathways. The altered interfaces between dikes and host rocks further function as preferential pathways for gas migration. The geometry and orientation of folds critically determine H_2 migration patterns within folded strata. Vertically oriented or steeply dipping folds typically exhibit enhanced H_2 channelling efficiency, whereas intricate fold architectures may promote gas dissipation. During metamorphism, metamorphic textures formed by geological processes may establish secondary migration channels or impermeable barriers to H_2 flow. Metamorphic processes, particularly near contact zones or within altered lithologies, exert a primary control on permeability evolution and gas migration dynamics [5, 12].

Fault systems act as primary conduits for subsurface hydrogen transport, with deep regional faults and shear zones channeling H_2 toward the surface, while secondary fracture networks locally regulate migration, especially near igneous intrusions or folded strata. Shallow faults often disperse H_2 accumulations due to their geometric complexity, whereas active tectonic regions enhance H_2 flux through dynamic fracture generation.

Intrusions, particularly discordant bodies, further influence H_2 distribution by creating preferential pathways along thermally and chemically altered contact zones, which facilitate fluid mobility. This is exemplified by H_2 migration documented adjacent to dike systems in the Lonnavale 1 well [12].

5.2 Accumulation

Natural hydrogen does not form laterally extensive, basin-scale accumulations comparable to conventional hydrocarbon systems. Instead, its occurrence is typically spatially discontinuous and structurally focused, reflecting fundamental differences in source mechanisms, migration behavior, and storage conditions. Whereas oil and gas accumulation depends on prolonged organic maturation within sedimentary basins and the preservation of buoyant fluids by regionally continuous traps, natural hydrogen is generated predominantly within basement environments and redistributed through tectonically controlled fault pathways.

Natural hydrogen accumulation fundamentally relies on the coupling between fault systems and hydrogen-generating basement terranes encompass a wide range of lithological domains (addressed in Chap. 4, including ophiolites, ultramafic and mafic peridotites, ancient cratonic basement, alkaline granites, and large igneous provinces). Once hydrogen molecules are generated, they preferentially migrate along fault zones, which act as high-efficiency fast pathways connecting deep hydrogen-generating environments with shallower levels of the crust.

During the migration along active faults, a portion of natural hydrogen leaks and accumulates at the surface, producing circular surface depressions known as 'fairy circles', as documented in multiple regions and discussed in Chap. 3. Beyond surface leakage, natural hydrogen can also accumulate in the shallow subsurface following fault-guided migration pathways, with reservoirs commonly located at depths of less than ~2 km. These reservoirs typically consist of multi-layered fracture–karst carbonate systems, or composite reservoir assemblages associated with doleritic sills or basaltic dikes, in which caprocks and aquifers jointly contribute to hydrogen sealing and storage, as exemplified by the Mali Bourakébougou hydrogen field [6]. In addition, transient hydrogen pooling may occur within fault zones where efficient hydrogen flow is focused along fault planes. If dynamic replenishment from hydrogen-generating source rocks persists within the fault system, a cyclic charging mechanism can develop, sustaining hydrogen supply over time, as proposed for the Bulqizë ophiolite).

The accumulation of H_2 occurs in three distinct phases: free, inclusion-bound, and dissolved gas. (i) Free hydrogen is distributed across multiple geological formations, including rock matrices, ore deposits, and fault zones. Research shows this mobile H_2 can traverse significant distances, enabling its accumulation near surface environments [13]. (ii) Inclusion-bound hydrogen is typically concentrated at greater depths, trapped within mineral inclusions or crystalline structures, and depends on tectonic activity for release into free-phase reservoirs [14]. (iii) Dissolved hydrogen predominantly resides in deep subsurface fluids under low-pressure conditions. Its behavior is marked by progressive exsolution, forming gas bubbles as carrier fluids migrate upward through fractures or permeable conduits [15].

The occurrence of free hydrogen in shallow geological layers indicates a greater potential for accumulation, supporting the hypothesis that regions with elevated free hydrogen concentrations may signal more substantial natural hydrogen reserves [5]. The presence of abundant fractures in the deeper geological strata facilitates the upward migration of H_2 into the shallow primary reservoir, whereas the overlying shallow doleritic sill, which is largely devoid of fractures, acts as an effective seal.

Similar to conventional hydrocarbon systems, the accumulation and retention of natural hydrogen are fundamentally governed by lithological and structural characteristics, including porosity, permeability, and caprock integrity, which collectively dictate migration pathways and storage mechanisms [16]. Petrophysical properties and lithology exert primary control over subsurface hydrogen behavior, wherein unconsolidated or highly fractured rocks facilitate H_2 mobility, while low-permeability formations such as intrusive bodies or evaporites may act as temporary caprocks, forming transient traps. However, due to its high volatility and the absence of sustained replenishment, accumulated H_2 often depletes rapidly upon extraction, as evidenced by wells such as Mt Kitty 1 and Sue Duroche #2. Although impermeable barriers can retard migration and prolong retention, such accumulations are generally unsuitable for long-term exploitation without ongoing subsurface generation or recharge. Macroscopic tectonic features (e.g., faults and shear zones) and microscopic textural attributes (e.g., cleavage and bedding planes) further focus fluid flow through structurally weakened conduits, enabling local enrichment while also potentially promoting dissipation through fracture networks [6].

5.3 Hydrogen Transport Mechanisms in Different Geological Environments

5.3.1 Migration of Free Hydrogen

Free hydrogen refers to molecular H_2 that occurs in the pores or fractures of rocks (or strata) and is capable of free migration [18–23]. As a principal form of natural hydrogen, it commonly occurs at concentrations ranging from 2 to 90%. The migration of free hydrogen within rock pores or fractures is driven by pressure differentials, concentration gradients, and hydrodynamic forces. Enhanced migration phenomena are particularly pronounced in structurally active geological settings such as fault zones and fold belts, where favorable pathways for H_2 transport are well-developed [24].

The earliest discovery of persistent gas seeps with continuous combustion was documented near Antalya, Turkey, where active vents have existed for over 2500 years

[25]. Approximately two centuries ago, another similar seepage site was identified in the Philippines, where flames generated by gas combustion have persisted to the present day, maintaining stable H_2 concentrations measured between 41.4% and 44.5% [26]. Current research reveals that natural hydrogen has been detected in various geological settings across more than 30 countries, including ophiolites, Precambrian rocks, igneous formations, volcanic gas emissions, hydrothermal systems (geysers and hot springs), tubular kimberlite bodies, ore deposits, oil–gas fields, coal basins, sedimentary sequences, and rock salt formations. Notably, free hydrogen concentrations exhibit significant variations depending on geological environments, specific generation zones within similar geological settings, and different sampling depths. Measured values range from single-digit percentages to exceptionally high concentrations exceeding 90% [27].

5.3.2 Migration of Inclusion-hosted Hydrogen

Inclusion-hosted hydrogen refers to H_2 trapped within rocks in the form of fluid inclusions or adsorbed phases, with concentrations ranging from 0.2–100% [23]. Analytical studies have revealed that H_2 predominantly exists as a confined gas phase in various lithologies, either encapsulated as fluid inclusions or adsorbed onto mineral surfaces. Similar to free hydrogen occurrences, molecular hydrogen has been identified in inclusion samples from diverse geological settings across more than ten countries, including ultramafic rocks, Precambrian basement formations, igneous complexes, volcanic sequences, kimberlite pipes, ore bodies, coal-bearing basins, sedimentary strata, metamorphic terrains, and evaporite deposits [27].

The analytical results of the samples demonstrate that the H_2 concentration within fluid inclusions exhibits significant variability, ranging from 0.2–100%. The migration mechanisms of inclusion-hosted H_2 are relatively complex, generally requiring the involvement of geological processes (such as tectonic movements, hydrothermal activities, etc.) to facilitate its release and subsequent migration. During processes such as rock fracturing, deformation, or fluid circulation, encapsulated H_2 is liberated from its primary host and subsequently transported by fluid phases to new storage spaces through migration pathways.

5.3.3 Migration of as Dissolved Hydrogen

Dissolved hydrogen denotes naturally occurring hydrogen gas that is found in groundwater in a dissolved state. Its concentration varies from trace amounts to several tens of percent. Dissolved hydrogen primarily migrates through groundwater flow [23]. Within groundwater systems, the transport of dissolved hydrogen is influenced by multiple factors, including flow velocity, flow direction, temperature, pressure, and properties of geological media. Under specific conditions (e.g., fluctuations in groundwater levels, changes in groundwater temperature), the concentration and spatial distribution of dissolved hydrogen may undergo alterations [28]. Natural hydrogen has been widely documented in numerous studies as occurring in the dissolved phase within groundwater. For instance, in a Soviet-era study on H_2 occurrence in subsurface fluids, researchers compiled a H_2 anomaly distribution map based on over 2000 analytical results, revealing higher anomaly values in regions associated with deep-seated faults and rift zones. H_2 has also been detected in water samples from over ten countries and in oil–gas field brines, measured concentrations in groundwater span from trace amounts to several tens of percent by volume.

Studies have shown that hydrogen solubility exhibits a positive correlation with pressure, whereas it declines as temperature or groundwater salinity increases. Under standard atmospheric pressure at 298 K (24.85°C), the solubility of molecular hydrogen in pure water is on the order of 7.9×10^{-4} mol/kg, while methane and carbon dioxide exhibit higher solubilities of 1.4×10^{-3} mol/kg and 3.3×10^{-2} mol/kg [29], respectively, under identical conditions. Consequently, H_2 demonstrates exceptionally low aqueous solubility as a non-polar gas compared to methane (CH_4) and carbon dioxide (CO_2). Investigations of H_2 content in oil–gas field formation waters reveal that H_2 accumulation levels in these systems exceed those observed in typical groundwater samples, approaching concentrations characteristic of natural hydrogen system [27]. This suggests potential genetic linkages between hydrocarbon generation processes and natural hydrogen formation mechanisms. Furthermore, H_2 serves as a sensitive redox indicator, exhibiting positive correlations with contaminant formation. Global observations indicate dissolved natural hydrogen concentrations exceeding 10% in groundwater systems at multiple locations [27]. The measurement of dissolved H_2 in subsurface waters has been effectively employed as an environmental monitoring tool for detecting contamination from landfill leachate, chemical solvents, and petroleum products in shallow aquifers [30, 31].

Building on the accumulation mechanisms outlined above, recent studies increasingly indicate that ultramafic units located beneath conventional reservoir horizons can function as intrinsic hydrogen-holding systems, rather than merely serving as source rocks. Rather than being confined to a single trapping mode, hydrogen behavior within these ultramafic systems reflects a dynamic interplay between generation, migration, retention, and release processes. Hydrogen is primarily generated through serpentinization reactions involving olivine-rich peridotites. Once produced, hydrogen migrates through the surrounding rock mass via free hydrogen diffusion, inclusion-hosted hydrogen or dissolved hydrogen, allowing it to migrate away from localized reaction sites. During this migration process, multiple retention pathways may operate simultaneously. Hydrogen molecules can become adsorbed onto serpentine mineral surfaces, be capillarily trapped as confined gas bubbles within micro- to nano-scale pore spaces, or dissolve into circulating groundwater systems.

The relative importance of these storage states varies with local geological conditions, including pressure, temperature, pore-fracture structure, and fluid chemistry. Hydrogen may transition between adsorbed, inclusion-hosted, dissolved, and free-gas phases, with desorption or pressure reduction enabling remobilization into the free-gas state. As a result, hydrogen accumulation within peridotitic reservoirs is inherently transient and dynamic, rather than static.

Ultimately, a portion of the generated hydrogen escapes the subsurface system and manifests as surface seepage, whereas another fraction is retained within peridotite-hosted reservoirs, contributing to subsurface natural hydrogen enrichment. This dual outcome highlights the role of ultramafic reservoirs not only as hydrogen sources but also as temporary to semi-stable storage media within natural hydrogen systems.

At deep earth, natural hydrogen behaves as a hybrid geochemical–physical system. While its generation and migration are strongly influenced by rock-fluid reactions, its subsurface transport is largely governed by physical processes, which are dominated by diffusion mechanism and flow mechanics in porous media. Importantly, hydrogen generation, migration, and leakage are not sequential but concurrent, forming a continuously evolving system in which supply and loss operate in parallel. As a result, natural hydrogen does not accumulate as a stable or static resource like conventional hydrocarbon; instead, it exists within a fundamentally non-equilibrium regime characterized by transient fluxes and spatially heterogeneous, discontinuous distributions. This dynamic behavior fundamentally distinguishes natural hydrogen from conventional hydrocarbons and necessitates a paradigm shift in exploration and development strategies, favoring high-resolution, time-dependent monitoring, adaptive assessment frameworks, and real-time reservoir evaluation rather than traditional static resource appraisal approaches.

References

1. Zou, J., Rezaee, R., and Yuan, Y. 2018. Investigation on the adsorption kinetics and diffusion of methane in shale samples. *Journal of Petroleum Science and Engineering* 171: 951–958.
2. Zou, J., et al. 2020. Distribution of adsorbed water in shale: An experimental study on isolated kerogen and bulk shale samples. *Journal of Petroleum Science and Engineering* 187: 106858.
3. Lodhia, B.H., L. Peeters, and E. Frery. 2024. A review of the migration of hydrogen from the planetary to basin scale. *Journal of Geophysical Research: Solid Earth* 129 (6): e2024JB028715
4. Yuan, Y., et al. 2023. High-resolution coupling of stratigraphic 'sweet-spot' lithofacies and petrophysical properties: A multiscale study of Ordovician Goldwyer Formation, Western Australia. *Petroleum Science* 20:1312–1326
5. Wei, Q., et al. 2023. Geological characteristics, formation distribution and resource prospects of natural hydrogen reservoir. *Natural Gas Geoscience* 35 (6): 1113–1122 (in Chinese).
6. Maiga, O., et al. 2023. Characterization of the spontaneously recharging natural hydrogen reservoirs of Bourakebougou in Mali. *Scientific reports* 13 (1): 11876–11876.
7. Maiga, O., et al. 2024. Trapping processes of large volumes of natural hydrogen in the subsurface: The emblematic case of the Bourakebougou H_2 field in Mali. *International Journal of Hydrogen Energy* 50: 640–647.

8. Guo, Z. 2001. Hydrocarbon-bearing prospects of volcanic rock covered regions in the southeastern coastal waters of China judged by the distribution of global oil and gas fields. *Petroleum Geology & Experiment* 2001 (02): 122–132 (in Chinese).
9. Meng, Q., et al. 2021. Geological background and exploration prospects for the occurrence of high-content hydrogen. *Petroleum Geology & Experiment* 43 (02): 208–216 (in Chinese).
10. Wan, J., et al. 2025. Lithofacies classification and reservoir property of lacustrine shale, the Cretaceous Qingshankou formation, Songliao basin, northeast China. *Marine and Petroleum Geology* 173: 107262.
11. Wan, J., et al. 2025. Laminae in multiple lithofacies and impact on pore structures in lacustrine shale: the Cretaceous Qingshankou Formation, Songliao Basin. *Marine and Petrole-um Geology* 107321.
12. Prinzhofer A, Cissé CST, Diallo AB. 2018. Discovery of a large accumulation of natural hydrogen in Bourakebougou (Mali). International Journal of Hydrogen Energy: 43(42):19315-26
13. Abe, J. O., et al. 2019. Hydrogen energy, economy and storage: Review and recommendation. *International Journal of Hydrogen Energy* 44 (29): 15072–15086.
14. Hematpur, H., et al. 2022. Review of underground hydrogen storage: Concepts and challenges. *Advances in Geo-Energy Research* 7: 111–131.
15. Zivar, D., S. Kumar, and J. Foroozesh. 2021. Underground hydrogen storage: A comprehensive review. *International Journal of Hydrogen Energy* 46 (45): 23436–23462.
16. Yuan, Y. 2020. Multi-Scale Porosity and Pore Structure Assessment of Shale. Curtin University.
17. Boreham, C.J., et al. 2021. Hydrogen in Australian natural gas: Occurrences, sources and resources. *The Australian Energy Producers Journal* 61 (1): 163–191.
18. Şimşek, E., O. Parlak, and A. H. F. Robertson. 2023. Ion-probe (SIMS) U-Pb geochronology and geochemistry of the Upper Cretaceous Kızıldağ (Hatay) ophiolite: Implications for supra-subduction zone spreading in the Southern Neotethys. *Geosystems and Geoenvironment* 2 (3): 100165.
19. Wang, L., et al. 2023. The origin and occurrence of natural hydrogen. *Energies* 16 (5): 2400–2400.
20. Bourdet, J., et al. 2023. Natural hydrogen in low temperature geofluids in a Precambrian granite, South Australia. Implications for hydrogen generation and movement in the upper crust. *Chemical Geology* 638.
21. Sleep, N. H., et al. 2004. H_2-rich fluids from serpentinization: Geochemical and biotic implications. *Proceedings of the National Academy of Sciences of the United States of America* 101 (35): 12818–12823.
22. Allen, D. E., and W. E. Seyfried. 2004. Serpentinization and heat generation: Constraints from Lost City and Rainbow hydrothermal systems. *Geochimica et Cosmochimica Acta* 68 (6): 1347–1354.
23. Li, Q., et al. 2024. Occurrence states and transformation of natural hydrogen mechanism. *Geological Survey of China* 11 (05): 9–20 (in Chinese).
24. Ballentine, C. J., and P. G. Burnard. 2002. Production, release and transport of noble gases in the continental crust. *Reviews in Mineralogy & Geochemistry* 47 (1): 481–538.
25. Etiope, G. 2023. Massive release of natural hydrogen from a geological seep (Chimaera, Turkey): Gas advection as a proxy of subsurface gas migration and pressurised accumulations. *International Journal of Hydrogen Energy* 48 (25): 9172–9184.
26. Abrajano, T. A., et al. 1988. Methane-hydrogen gas seeps, Zambales Ophiolite, Philippines: Deep or shallow origin? *Chemical Geology* 71 (1): 211–222.
27. Zgonnik, V. 2020. The occurrence and geoscience of natural hydrogen: A comprehensive review. *Earth-Science Reviews* 203: 103140.

28. Ballentine, C. J., et al. 2025. Natural hydrogen resource accumulation in the continental crust. *Nature Reviews Earth & Environment* 6 (5): 342–356.
29. Duan, Z. H., and R. Sun. 2003. An improved model calculating CO_2 solubility in pure water and aqueous NaCl solutions from 273 to 533 K and from 0 to 2000 bar. *Chemical Geology* 193 (3–4): 257–271.
30. Bjerg, P. L., et al. 1997. Effects of sampling well construction on H_2 measurements made for characterization of redox conditions in a contaminated aquifer. *Environmental Science & Technology* 31 (10): 3029–3031.
31. Chapelle, F. H., et al. 1996. Comparison of Eh and H_2 measurements for delineating redox processes in a contaminated aquifer. *Environmental Science & Technology* 30 (12): 3565–3569.

6 Hydrogen Storage and Mechanism

The mechanisms governing H_2 storage share similarities with those observed in conventional oil and gas systems [1, 2]. Effective subsurface H_2 storage is enabled by impermeable caprocks and structural traps, which act as barriers to inhibit vertical migration, allowing accumulations to form over geological timescales [3]. Prolonged H_2 leakage observed in surface features such as "fairy circles" provides evidence for dynamic reservoir processes, where transient H_2 reserves are replenished by ongoing geological activity. Research indicates a depth-dependent increase in H_2 concentration, suggesting these reserves are responsive to subsurface changes rather than static accumulations [4]. The complex behavior of natural hydrogen systems, governed by factors such as microbial activity, rock-fluid interactions, and tectonic shifts, underscores the necessity for advanced modelling and targeted exploration to quantify their sustainable potential [5].

In the Bourakebougou field, the doleritic rock constitutes the principal trapping mechanism [6]. Both the upper and lower doleritic layers function as seals, with the upper layer exhibiting marginally reduced porosity, thereby enhancing its sealing efficacy. The upper unit extends across a thickness of 15–50 m, though its sealing performance declines significantly below 20 m depth. The seal is predominantly constituted by plagioclase and pyroxene, with accessory biotite and opaque minerals, collectively forming igneous lithologies characterized by low porosity. Increased porosity at specific depths correlates with heightened fracturing within the dolerite. Furthermore, a silty-shale interval is observed beneath the primary dolerite caprock in multiple wells, suggesting a secondary sealing component in the system [6].

To investigate the H_2 trapping potential of the silty shale layer beneath the primary dolerite unit, thickness mapping and mineralogical analyses can be performed [6]. The findings revealed an inverse correlation between the thickness of this layer, conventionally

Y. Yuan et al., *Natural Hydrogen*, Synthesis Lectures on Renewable Energy Technologies, https://doi.org/10.1007/978-3-032-14469-0_6

recognized for oil and gas entrapment, and its H_2 retention capacity. A similar inverse trend was observed in the Breccia layers. Mineralogical assessment indicated that the layer was composed mainly of a shaly matrix and quartz grains, with minimal clay content [5, 7].

To achieve geological storage of H_2, it is essential to ensure sufficient storage capacity and sealing integrity [8]. This requires reservoirs possessing both high storage space to accommodate the required gas volume and high permeability to enable adequate rates of gas injection and withdrawal. Among common geological settings, porous reservoir rocks, including depleted hydrocarbon fields and saline aquifers, are often overlain by caprocks that provide an effective seal for gas containment [9, 10]. Certain mined caverns (e.g., salt caverns) also serve as viable storage facilities due to the effective sealing properties of their surrounding rock formations. Aquifers, depleted hydrocarbon reservoirs, and salt caverns constitute the three principal options for underground hydrogen storage [11].

The geological reservoirs suitable for H_2 storage include basalt formations, organic-rich shales, tight gas reservoirs, depleted oil and gas reservoirs, saline aquifers, and coal seams [12, 13]. Among these, the interfacial tension between rock and fluid phases and the wetting behavior of rocks in the presence of H_2 constitute critical parameters governing H_2 immobilization in subsurface environments [14, 15]. Wettability fundamentally determines the spatial distribution of H_2 during its migration through microchannels within porous rock matrices. The wettability characteristics shaped by geological formations and associated interfacial phenomena represent key factors modulating H_2 propagation capacity throughout reservoir strata [16[. These properties collectively govern gas injectivity, withdrawal rates, hydrodynamic flow behavior, and storage potential while reducing operational uncertainties [17]. In such scenarios, caprocks provide essential sealing capacity to prevent upward H_2 migration, thereby enabling permanent sequestration within geological structures [11].

Natural hydrogen system exhibits five defining geological attributes [18]. These features include: (1) broad spatial distribution across diverse geological settings, encompassing continental and oceanic crusts as well as volcanic-hydrothermal systems; (2) fault and fracture networks serving as dominant conduits for H_2 migration; (3) subsurface free hydrogen abundance governing the scale of the resource, serving as a critical parameter for reserve estimation; (4) highly enriched zones (often termed "sweet spots") typically located at shallow depths within reservoirs; and (5) dynamic storage mechanisms wherein H_2 flux and retention are continually modulated by ongoing geological processes. Collectively, these attributes establish a conceptual framework for analyzing natural hydrogen system dynamics, which originate from both biotic and abiotic processes.

6.1 Hydrogen Storage in Rock Body

Rocks represent one of the principal reservoirs for H_2 storage. Within the intricate architecture of porous lithologies such as sandstone, shale, and limestone, the multiscale pore network constitutes a microscopic domain for H_2 sequestration [19–22]. These pores can effectively store significant amounts of H_2 due to their remarkable specific surface area [23]. This critical petrophysical property enables extensive exposure of pore surfaces, thereby enhancing interfacial interactions with H_2 molecules and facilitating efficient adsorption and storage mechanisms [24–26]. The adsorption capacity of porous rocks primarily originates from physical and chemical adsorption sites distributed across pore surfaces. When H_2 molecules interact with these active sites, they are trapped primarily due to various intermolecular attractive forces, such as van der Waals forces and electrostatic interactions. This adsorption phenomenon demonstrates not only exceptional stability but also exhibits considerable resistance to external environmental perturbations [27, 28]. The combination of high surface reactivity and nanoconfined space effects in rock pores creates an optimized microenvironment for H_2 retention, where molecular-scale interactions dominate the storage dynamics.

Studies have identified H_2 as the dominant component within the trapped gas accumulation (occurring either as fluid inclusions or in adsorbed form) across various rock types. Each lithology exhibits a distinctive gas composition when liberated [29, 30]. However, quantitative inconsistencies persist in H_2 extraction records from fluid inclusions of rocks sharing identical sedimentary environments.

While numerous studies have documented the detection of H_2 as free gas in ophiolite contexts, only one recent investigation has identified H_2 analysis within fluid inclusions from such rocks (demonstrating both trapped and free-state hydrogen in ophiolites). Research reveals that H_2 constitutes the predominant component within fluid inclusions of olivine samples from the Zambales ophiolite, Philippines [31]. Recent analyses in Oman have revealed hydrogen-enriched fluid inclusions within ophiolitic rocks [32]. This finding underscores the potential underestimation of H_2 occurrence in lithospheric systems and emphasizes the necessity for systematic H_2 screening in fluid inclusion studies of ultramafic sequences.

Current literature documents two H_2 samples extracted from fluid inclusions within ultramafic rock formations. The origin of this H_2 is attributed to water–rock interactions, particularly serpentinization or other hydration-related geochemical processes (see Sect. 2.10 on serpentinization).

Hydrogen-rich gas inclusions have been documented in minerals from rift zone environments. For instance, studies of oceanic fissures report H_2 concentrations within these inclusions averaging as high as 21.4% [33]. However, systematic investigations targeting H_2-bearing inclusions in cratonic rift zones remain absent from existing literature.

Studies have documented elevated H_2 levels detected in fluid inclusions hosted within Precambrian lithologies. Aligning with free gas analyses, H_2 concentrations display a gradual increase with depth, peaking at maximum concentrations in Precambrian basement-hosted fluid inclusions [34]. Recent global analyses of Precambrian samples reveal that H_2 persists as a ubiquitous constituent, with an average concentration of 0.14% (by volume), whereas younger basement lithologies exhibit concentrations typically an order of magnitude lower [35].

According to reports, significant amounts of H_2 have been detected in gas inclusions within rocks of igneous origin. These examples challenge the assertion made in a previous study that 'water-saturated granites under crustal conditions are capable of retaining only trace quantities of H_2' [36]. In fact, another study reported exceptionally high H_2 concentrations reaching up to 99% in granite inclusions [37].

H_2 is commonly found trapped within gas inclusions of volcanic rock specimens. Research has shown that fresh volcanic rocks release H_2 at an average concentration of 45% during degassing processes [38]. Studies indicate that gas inclusions within kimberlite pipe samples exhibit elevated H_2 concentrations. For instance, analyses of fluid inclusions from the Obnazhennaya kimberlite pipe (Russia) reveal that H_2 dominates the composition of deep-seated fluids at depths greater than 70 km [37]. H_2 is also persistently identified as the principal gaseous constituent within diamond-hosted inclusions [39, 40]. Further investigations corroborate a depth-dependent increase in H_2 abundance, as evidenced by hydrogen-to-water ratio measurements in diamond inclusion gases. Notably, diamonds crystallizing at approximately 400 km depth display H_2 concentrations up to 140 times greater than those formed at shallower depths (~ 120 km) [41].

H_2 has been detected within fluid inclusions across diverse mineral deposits and ore systems, including iron, uranium, gold, nickel, chromium, and polymetallic formations. Analyses reveal that hydrogen (H_2) concentrations within fluid inclusions in quartz from gold–tungsten deposits exhibit a systematic increase with depth [42]. A substantial body of research has reported the ubiquitous occurrence of H_2 within gold-bearing deposit systems, with concentrations in gold-bearing veins reaching 200–460 cm^3/kg rock, marking a 100-fold enrichment relative to surrounding country rocks [43]. Subsequent analyses by the same researcher identified H_2-bearing inclusions in mercury ores containing H_2 concentrations up to 144 cm^3/kg rock, representing 19% of inclusion gases in certain cases [43]. Analysis of gas inclusions within silver-polymetallic hydrothermal minerals from the Kurusaisk mining district, Uzbekistan, revealed hydrogen concentrations of up to 42.6% (102 cm^3/kg) in quartz samples. Other minerals displayed the following gas compositions (percentage/(cm^3/kg)): garnet 23/36, sphalerite 12/15, calcite 4/1.9, galena 11.6/2.6, and barite 7.1/6.8 [44]. Elevated H_2 concentrations have been documented as trapped gases in rock samples from the Tyrnyauz tungsten-molybdenum deposit (Caucasus Mountains, Russia) [45]. Notably, a borehole located within the Tyrnyauz mine discharged hydrogen-rich gas continuously over a period of several months.

Numerous studies have examined the spatial distribution and chemical composition of geological gases within the Khibiny and Lovozero massifs on the Kola Peninsula in Russia. Studies indicate that H_2 concentrations peak in rocks adjacent to apatite-nepheline mineralized zones within these ore-rich regions. Gas accumulation occurs preferentially in anhydrous, highly microfractured rock blocks, which are bounded by low-permeability lithological units [46]. Analyses of pore systems reveal that closed pores in the Khibiny massif host the majority of trapped gases, while open pores exhibit restricted gas content [47]. Comparative analyses of matrix-hosted hydrogen and hydrocarbon gases from both massifs, when contrasted with sealed gases in fluid inclusions and free gases in fractures, demonstrate that H_2 constitutes the predominant dispersed gas in unperturbed geological contexts [48]. In the Khibiny massif, sealed gases contain average H_2 concentrations of 4%, whereas fracture gases show substantially elevated levels, averaging 20% [49]. Notably, H_2 concentrations in sealed gases exhibit a marked reduction with increasing depth [46]. No significant correlation was observed between the CH_4/H_2 ratios of free and sealed gases, nor between hydrogen and hydrocarbon concentrations within these systems [46, 50]. A recent synthesis comprehensively reviews regional gas distribution patterns and mechanisms [51].

H_2 concentrations were quantified in gas inclusions present in samples collected from multiple coal-bearing basins. Documented instances of H_2 occurrence within sedimentary or metamorphic rock formations remain relatively limited in published studies. This paucity could stem from the typically elevated porosity and permeability characteristics of sedimentary rocks, which promote H_2 migration and dispersion. However, controlled laboratory investigations of H_2 saturation in clastic and carbonate lithologies reveal substantial H_2 adsorption capacities (24–57 times baseline levels) with multi-day retention stability, indicating promising reservoir characteristics through combined adsorbed and free-phase storage mechanisms comparable to conventional natural gas systems. Among examined lithologies, limestone demonstrates superior adsorption performance [52]. Systematic laboratory measurements of H_2 concentrations across diverse sedimentary formations show measurable quantities in the majority of specimens. Carbonate rocks exhibit peak values up to 77.2 cm^3/kg (average 14.8 cm^3/kg), implying an intrinsic propensity for H_2 sequestration in carbonate matrices [53]. Exceptional H_2 accumulations (20–30 vol%) have been recorded in anhydrite/gypsum-bearing formations [54]. Modern experimental simulations assessing sandstone's viability as H_2 reservoir media under H_2-saturated conditions demonstrated reactive modifications of pore-lining anhydrite and carbonate cementation [55]. H_2 manifestations have additionally been detected within carbonaceous shale units and metamorphic terranes [56]. Cutting-edge investigations of argillaceous sequences in the Cigar Lake uranium deposit reveal extraordinary H_2 enrichment levels (up to 0.25 mol/kg), surpassing standard methane adsorption capacities observed in shale reservoirs [57].

Previous studies have documented the detection of H_2 in fluid inclusions trapped within salt deposits, as evidenced by analytical data from salt samples. Researchers have proposed that H_2 enrichment in such geological formations may result from the lack of reactive chemical species capable of reacting with H_2, enabling its preservation over time

[54]. Notably, gas inclusions analyzed from salt mines in Hesse and Thuringia, Germany, confirmed widespread H_2 occurrence, with peak concentrations recorded at 32.9% in Thuringian geological specimens and 26.4% in samples from Hesse.

6.2 Hydrogen Storage in Fracture and Fault

Structural spaces such as fractures and faults serve as important reservoirs for H_2 storage. These geological features typically exhibit high connectivity and permeability, which enable them to store significant amounts of H_2. Additionally, fractures and fault systems can function as effective migration conduits, facilitating H_2 transportation and accumulation between different subsurface strata [58].

Numerous studies have demonstrated the association between H_2 and fault zones, as these geological structures serve as natural pathways for H_2 migration. The first experimental study conducted in Japan revealed a significant correlation between H_2 concentrations and both fault presence and seismic activity [59]. Subsequent investigations documented exceptionally high H_2 concentrations reaching 9.36% in specific Japanese fault systems [60]. These findings collectively indicate that H_2 detection can serve as an effective tool for fault mapping and characterization [61].

Studies on H_2n in soil gases have been conducted along the San Andreas Fault (California) and Dixie Valley Fault System (Utah), USA. A pronounced correlation has been observed between H_2 and helium concentrations in soil gases and underlying deep-seated fault zones. Elevated helium concentrations reaching 1243 ppm in certain areas provide indirect evidence for the geological origin of H_2 when associated with helium anomalies [62]. Geochemical H_2 surveying in the United States has proven effective in delineating subsurface fracture networks, notably along Idaho's Trans-Challis Fault System and near mineralized structures within Nevada's Carlin Trend. Researchers have proposed that systematic analysis of H_2 anomalies could serve as an effective tool for fault zone mapping [63]. Based on the estimated global H_2 flux of 2.7×10^4 t/yr [64], the calculated H_2 migration rate along fault zones was determined as 3.6×10^{-7} $cm^3 \cdot s$ [65]. This value represents a conservative estimate compared with other published calculations.

Analysis of soil gas from numerous sampling locations throughout north-central Kansas indicates that fractures serve as primary conduits for the vertical migration of molecular hydrogen [66]. Regional synthesis of H_2 distribution in Ukraine and Belarus demonstrated that subsurface fluids within the North Pryp'yat Tectonic Zone contain the most significant H_2 accumulations recorded in these regions [67]. Evaluation of over 2000 groundwater H_2 measurements further established that waters intersecting deep-seated faults exhibit H_2 concentrations up to 100 times greater than those from non-tectonic settings [68].

A geophysical study analyzing deep-seated fault systems within the Moscow Region, Russian Federation, integrated subsurface H_2 measurements and demonstrated a clear relationship between peak H_2 concentrations and these tectonic structures [69]. Complementary research on kimberlite pipes revealed heightened concentrations of H_2-rich gases spatially aligned with pervasive fracture networks, suggesting these features act as preferential pathways for subsurface gas migration [70].

Subsurface hydrogen concentrations were comprehensively measured throughout prospective geological domains in southwestern Belarus. The findings reveal that significantly elevated H_2 concentrations are structurally controlled by deep-seated fault systems. In these zones, H_2 levels exceed background concentrations by factors of 10–50, with background values closely corresponding to atmospheric H_2 concentrations. A statistically significant correlation ($r = 0.7$) between H_2 and He concentrations indicates a common genetic link associated with deep crustal fault systems [71, 72].

Structural discontinuities are critical in regulating the migration of mantle-derived fluids into sedimentary basins. A notable case study from the North Sea hydrocarbon reservoirs employs multi-isotopic tracers (Os, C, He, Ar, Ne, O, and D) to establish fluid origins, identifying a substantial mantle-derived contribution to accumulated hydrocarbons [73–75]. Although direct measurements of H_2 flux remain elusive, helium flux has been validated as a reliable analog for tracking H_2 migration, as evidenced in these investigations. This integrative methodology advances understanding of subsurface fluid dynamics and transport pathways in crustal environments.

6.3 Hydrogen Storage in Dissolved Groundwater

Dissolved hydrogen in groundwater is also a significant form of hydrogen storage. During its migration through subsurface environments, groundwater dissolves a portion of molecular hydrogen, forming hydrogen-enriched groundwater. These hydrogen-bearing aqueous solutions are stored and transported within the pore spaces of subsurface rock formations, thereby serving as a critical carrier for H_2 storage and migration processes [23]. This hydrogeological mechanism provides essential pathways for both the accumulation and lateral redistribution of H_2 resources in sedimentary basins.

Multiple occurrences of natural hydrogen dissolved in groundwater have been reported globally. During a systematic review of H_2 occurrences in subsurface fluids across former Soviet-bloc territories, a H_2 anomaly map was developed using data from more than 2,000 geochemical analyses. This study identified that anomalously high H_2 concentrations were primarily concentrated in regions marked by recent tectonic activity, where groundwater monitoring wells detected significant volumes of free H_2 [67]. A separate study analyzing 2215 H_2 measurements in groundwater reported background levels of 50 mL/L in stable cratonic regions, with anomalies exceeding 1500 mL/L near deep-seated faults and rift systems [68]. Statistical evaluation of dissolved H_2 in Western Siberian groundwater revealed detectable H_2 levels in 15% of collected samples, ranging from trace quantities to concentrations of several tens of percentage points. Elevated H_2 concentrations were detected with significantly greater frequency in deeper sampling intervals [76]. However, recent studies emphasize the lack of conclusive evidence or documented cases of natural springs enriched with H_2 [77]. The author highlights that H_2 detection is not routinely incorporated into standard analytical procedures and remains methodologically challenging, which may account for the limited discoveries thus far.

H_2 enrichment has been identified in groundwater systems hosted within fractured rock aquifers across 24 South African boreholes. Notably, dissolved H_2 levels exhibited no statistically significant correlations with depth, salinity, pH, lithology, borehole age, or aquifer age [78]. However, peak H_2 concentrations were predominantly associated with deeper, saline, non-meteoric groundwater sourced from older fractured aquifers. A distinct study in Idaho, USA, documented anomalously high H_2 levels in groundwater wells, with geochemical evidence implicating a Pliocene caldera as a potential source [79]. In Crimea, Ukraine, groundwater analyses revealed dissolved hydrogen concentrations spanning 0.2–53.6 vol%, with both concentration and total hydrogen content increasing tenfold at greater depths. Intriguingly, no clear relationship emerged between H_2 and dissolved iron concentrations in this region [80]. Contrastingly, groundwater in Russia's Volga River basin demonstrated a recurring spatial association between elevated H_2 levels, high iron concentrations (328 mg/L), and acidic pH conditions [81], highlighting divergent hydrogeochemical controls on H_2 occurrence.

The quantification of dissolved hydrogen in groundwater has emerged as an effective technique for assessing shallow aquifer systems impacted by contaminants such as landfill leachate [82], solvents, and jet fuels [83]. H_2 serves as a redox indicator demonstrating positive correlation with contaminant plumes. It has been recommended that monitoring wells should be permitted to equilibrate for up to three months before measurement implementation [82]. Furthermore, H_2 quantification has been demonstrated to provide more reliable identification of anaerobic redox processes compared to conventional Eh (redox potential) measurements [83].

6.4 Hydrogen Storage Mechanisms in Different Geological Environments

The mechanisms governing hydrogen storage across various geological settings are outlined as follows (Table 6.1).

Table 6.1 The mechanisms of hydrogen accumulation and storage in different geological environments, including igneous rocks, kimberlites, orebodies, precambrian Shields, salt deposits, sedimentary and metamorphic rocks, ultramafic rocks and coal basins

Geological environment	The mechanisms of hydrogen storage
Igneous rocks	Hydrogen may accumulate within pore spaces or cavities of igneous rock formations, where migration is impeded by low-permeability rock layers [84]
Kimberlites	Hydrogen tends to accumulate within fractures and porous geological formations proximal to kimberlite pipes. The host structures are encased within impermeable lithological strata [85]
Orebodies	Hydrogen preferentially accumulates within pore spaces and fractures proximal to ore bodies, particularly when enclosed by low-permeability lithologies
Precambrian shields	Hydrogen accumulations primarily form within fault zones and fractures where fluid migration is confined. The crystalline structure of shield rocks promotes the concentration of hydrogen within localized reservoirs
Salt deposits (evaporites)	Hydrogen concentrations predominantly form within fault systems and fracture networks where fluid migration is impeded
Sedimentary and metamorphic rocks	Hydrogen can accumulate in permeable sandstone reservoirs or within fractures and cavities generated by metamorphic processes. The impermeable caprocks improve their capacity to trap and preserve hydrogen [84]
Ultramafic rocks	Hydrogen accumulates within fracture networks and pore spaces in ultramafic rocks, which is controlled by the structural architecture of the host rock and the presence of impermeable confining strata [86]
Volcanic formations	Hydrogen becomes concentrated within permeable volcanic rock formations or below solidified lava deposits, where it is trapped by impermeable caprock layers
Coal basins	Hydrogen may accumulate either within the coal seams or in surrounding porous rock formations, where impermeable geological barriers restrict its migration [87]

References

1. Mahdi, D.S., et al. 2021. Hydrogen underground storage efficiency in a heterogeneous sandstone reservoir. *Advances in Geo-Energy Research* 5: 437–443.
2. Yuan, Y., et al. 2023. A comprehensive review on shale studies with emphasis on nuclear magnetic resonance (NMR) technique. *Gas Science and Engineering* 120: 205163.

3. Zivar, D., S. Kumar, and J. Foroozesh. 2021. Underground hydrogen storage: A comprehensive review. *International Journal of Hydrogen Energy* 46 (45): 23436–23462.
4. Hosseini, M., Experimental investigation of the interface and wetting characteristics of rock-H_2-brine systems for H_2 geological storage. 2023.
5. Tian, Q., et al. 2022. Origin, discovery, exploration and development status and prospect of global natural hydrogen under the background of "carbon neutrality." *China Geology* 5 (4): 722–733.
6. Maiga, O., et al. 2024. Trapping processes of large volumes of natural hydrogen in the subsurface: The emblematic case of the Bourakebougou H_2 field in Mali. *International Journal of Hydrogen Energy* 50: 640–647.
7. Prinzhofer, A., C. S. T. Cissé, and A. B. Diallo. 2018. Discovery of a large accumulation of natural hydrogen in Bourakebougou (Mali). *International Journal of Hydrogen Energy* 43 (42): 19315–19326.
8. Muhammed, N. S., et al. 2022. A review on underground hydrogen storage: Insight into geological sites, influencing factors and future outlook. *Energy Reports* 8: 461–499.
9. Krevor, S., et al. 2023. Subsurface carbon dioxide and hydrogen storage for a sustainable energy future. *Nature Reviews Earth & Environment* 4 (2): 102–118.
10. Xu, J., et al. 2021. New insights into controlling factors of pore evolution in organic-rich shale. *Energy & Fuels* 35(6): 4858–4873.
11. Wang, L., et al. 2024. Research progress in underground hydrogen storage (in Chinese). *Earth Science* 49 (06): 2044–2057.
12. Panfilov, M. 2010. Underground storage of hydrogen: In situ self-organisation and methane generation. *Transport in Porous Media* 85 (3): 841–865.
13. Ali, A., et al., Geochemical and C-O isotopic study of ophiolite-derived carbonates of the Barzaman Formation, Oman: Evidence of natural CO_2 sequestration via carbonation of ultramafic clasts. Journal of Geophysical Research-Solid Earth, 2021. 126(10).
14. Thaysen, E. M., et al. 2023. Pore-scale imaging of hydrogen displacement and trapping in porous media. *International Journal of Hydrogen Energy* 48 (8): 3091–3106.
15. Liu, J. Y., et al. 2023. Genesis and energy significance of natural hydrogen. *Unconventional Resources* 3: 176–182.
16. Yuan, Y., et al. 2021. Pore-scale study of the wetting behaviour in shale, isolated kerogen and pure clay. *Energy & Fuels* 35: 18459–18466.
17. Ellison, E.T., et al., Low-temperature hydrogen formation during aqueous alteration of serpentinized peridotite in the Samail ophiolite. Journal of Geophysical Research-Solid Earth, 2021. 126(6).
18. Wei, Q., et al. 2023. Geological characteristics, formation distribution and resource prospects of natural hydrogen reservoir (in Chinese). *Natural Gas Geoscience* 35 (6): 1113–1122.
19. Yuan, Y., et al. 2021. Compositional controls on nanopore structure of different shale lithofacies: A comparison with pure clays and isolated kerogens. *Fuel* 303: 121079.
20. Yuan, Y., and Rezaee, R., 2019. Fractal analysis of the pore structure for clay bound water and potential gas storage in shales based on NMR and N2 gas adsorption. *Journal of Petroleum Science and Engineering* 177: 756–765.
21. Yuan, Y., and Rezaee, R. 2019. Comparative porosity and pore structure assessment in shales: Measurement techniques, influencing factors and implications for reservoir characterization. *Energies* 12: 2094.
22. Yuan, Y., et al. 2018. Pore characterization and clay bound water assessment in shale with a combination of NMR and low-pressure nitrogen gas adsorption. *International Journal of Coal Geology* 194: 11–21.

23. Heinemann, N., et al. 2021. Enabling large-scale hydrogen storage in porous media - the scientific challenges. *Energy & Environmental Science* 14 (2): 853–864.
24. Yuan, Y., et al. 2019. Impact of composition on pore structure properties in shale: Implications for micro-/mesopore volume and surface area prediction. *Energy & Fuels* 33: 9619–9628.
25. Yuan, Y., and Rezaee, R. 2019. Impact of paramagnetic minerals on NMR-converted pore size distributions in Permian Carynginia Shales. *Energy & Fuels* 33: 2880–2887.
26. Yuan, Y., et al. 2018. Pore characterization and fluid distribution assessment of gas shale. Proceedings of 80th EAGE Conference and Exhibition, 1–5.
27. Gasanzade, F., et al. 2021. Subsurface renewable energy storage capacity for hydrogen, methane and compressed air – A performance assessment study from the North German Basin. *Renewable and Sustainable Energy Reviews* 149: 111422.
28. Войтов, Г. 1990. Восстановленные газы (углеводороды) в породах фундамента Русской плиты. *Бюллетень МОИП* 61: 44–61.
29. Giardini, A. A., G. V. Subbarayudu, and C. E. Melton. 1976. The emission of occluded gas from rocks as a function of stress: Its possible use as a tool for predicting earthquakes. *Geophysical Research Letters* 3 (6): 355–358.
30. Jiang, F., and G. Li. 1981. Experimental studies of the mechanisms of seismo-geochemical precursors. *Geophysical Research Letters* 8 (5): 473–476.
31. Grozeva, N.G., et al., Chemical and isotopic analyses of hydrocarbon-bearing fluid inclusions in olivine-rich rocks. Philosophical Transactions of the Royal Society a-Mathematical Physical and Engineering Sciences, 2020. 378(2165).
32. Zgonnik, V., et al. 2019. Diffused flow of molecular hydrogen through the Western Hajar mountains, Northern Oman. *Arabian Journal of Geosciences* 12 (3): 71.
33. Моисеенко, В. and В. Сахно, Глубинные флюиды, вулканизм и рудообразование тихоокеанского пояса. Наука, Москва., 1982.
34. Зорькин, Л., И. Старобинец, and Е. Стадник, Геохимия природных газов нефтегазоносных бассейнов. Москва, Недра, 1984.
35. Parnell, J. and N. Blamey, Hydrogen from radiolysis of aqueous fluid inclusions during diagenesis. Minerals, 2017. 7(8).
36. Apps, J. and P.C. Kamp, Energy gases of abiogenic origin in the Earth's crust. United States Geological Survey, Professional Paper; (United States), 1993. 1570.
37. Петерсилье, И, and В Припачкин. 1979. Водород, углерод и гелий в газах изверженных горных пород. *Геохимия* 7: 1026–1034.
38. Соколов, В., Газы земли. Наука, Москва, 1966.
39. Smith, E. M., et al. 2016. Large gem diamonds from metallic liquid in Earth's deep mantle. *Science* 354 (6318): 1403–1405.
40. Melton, C. E., and A. A. Giardini. 1974. The composition and significance of gas released from natural diamonds from Africa and Brazil. *American Mineralogist* 59 (7–8): 775–782.
41. Перчук, Л. 2000. Флюиды в нижней коре и верней мантии Земли. *Вестник Московского Университета* 4: 25–35.
42. Letnikov, F. A., and A. V. Narseev. 1991. Use of fluid inclusion gas surveys for the assessment of lode deposits (with reference to gold and tungsten deposits). *Journal of Geochemical Exploration* 42 (1): 133–142.
43. Фридман, А., Природные газы рудных месторождений. Недра, Москва, 1970.
44. Соколов, В., Геохимия природных газов. Недра, Москва, 1971.
45. Гуревич, М, et al. 1960. Материалы к геохимической характеристике природных газов рудных месторождений. *Труды Института Геологии Рудных Месторождений, Петрографии, Минералогии и Геохимии* 46: 83–91.

46. Нивин, В., Газовые компоненты в магматических породах: Геохимические, минералогические и экологические аспекты и следствия (на примере интрузивных комплексов Кольской провинции). 2013.
47. Онохин, Ф., Горючие газы Хибинского щелочного массива. Советская Геология, 1959: p. 109–118.
48. Nivin, V. A. 2009. Diffusively disseminated hydrogen-hydrocarbon gases in rocks of nepheline syenite complexes. *Geochemistry International* 47 (7): 672–691.
49. Nivin, V. A., et al. 2005. A review of the occurrence, form and origin of C-bearing species in the Khibiny alkaline igneous complex, Kola Peninsula. *NW Russia. Lithos* 85 (1–4): 93–112.
50. Нивин, В. 2001. О формах нахождения водородно-углеводородных газов в породах Хибинского и Ловозерского нефелин-сиенитовых массивов. *Щелочной магматизм Земли и его рудоносность* 2007: 200–203.
51. Nivin, V. A. 2019. Occurrence forms, composition, distribution, origin and potential hazard of natural hydrogen–hydrocarbon gases in ore deposits of the Khibiny and Lovozero massifs: A review. *Minerals* 9 (9): 535.
52. Levshounova, S. P. 1991. Hydrogen in petroleum geochemistry. *Terra Nova* 3 (6): 579–585.
53. Левшунова, С., О распространении сорбированного водорода в осадочных породах. Геология Нефти и Газа, 1982.
54. Smith, N. J. P., et al. 2005. Hydrogen exploration: A review of global hydrogen accumulations and implications for prospective areas in NW Europe. *Geological Society, London, Petroleum Geology Conference Series* 6 (1): 349–358.
55. Flesch, S., et al. 2018. Hydrogen underground storage-Petrographic and petrophysical variations in reservoir sandstones from laboratory experiments under simulated reservoir conditions. *International Journal of Hydrogen Energy* 43 (45): 20822–20835.
56. Suzuki, N., H. Saito, and T. Hoshino. 2017. Hydrogen gas of organic origin in shales and metapelites. *International Journal of Coal Geology* 173: 227–236.
57. Truche, L., et al. 2018. Clay minerals trap hydrogen in the Earth's crust: Evidence from the Cigar Lake uranium deposit. *Athabasca. Earth and Planetary Science Letters* 493: 186–197.
58. Ma, X., et al. 2022. Development directions of major scientific theories and technologies for underground gas storage (in Chinese). *Natural Gas Industry* 42 (05): 93–99.
59. Wakita, H., et al., Hydrogen release: new indicator of fault activity. Science (New York, N.Y.), 1980. 210(4466): p. 188–90.
60. Sugisaki, R., et al. 1983. Origin of hydrogen and carbon dioxide in fault gases and its relation to fault activity. *The Journal of Geology* 91 (3): 239–258.
61. Ware, R. H., C. Roecken, and M. Wyss. 1984. The detection and interpretation of hydrogen in fault gases. *Pure and applied geophysics* 122 (2): 392–402.
62. Jones, V.T. and R.J. Pirkle, Helium and hydrogen soil gas anomalies associated with deep or active faults. 1981.
63. McCarthy, J.H. and T.H. Kiilsgaard, Soil gas studies along the Trans-Challis fault system near Idaho City, Boise County, Idaho, in Bulletin. 2001.
64. Giardini, A. A., and C. E. Melton. 1983. A scientific explanation for the origin and location of petroleum accumulations. *Journal of Petroleum Geology* 6 (2): 117–138.
65. Su, Q., E. Zeller, and E. Angino. 1992. Inducing action of hydrogen migrating along faults on earthquakes (in Chinese). *Acta Seismologica Sinica* 5 (4): 841–847.
66. McCarthy Jr, J.H., et al., Soil gas studies around hydrogen-rich natural gas wells in northern Kansas, in Open-File Report. 1986.
67. Shcherbakov, A. V., and N. D. Kozlova. 1986. Occurrence of hydrogen in subsurface fluids andthe relationship of anomalous concentrations to deep faults in the USSR. *Geotectonics* 20: 120–128.

68. Щербаков, А., Проблема водородных подземных флюидов. Дегазация Земли и геотектоника, 1985: p. 164–165.
69. Rogozhin, E. A., et al. 2010. Deep structure of the Moscow aulacogene in the western part of Moscow. *Izvestiya Atmospheric and Oceanic Physics* 46 (8): 973–981.
70. Сороченко, М, and А Дроздов. 2010. Геохимические особенности природных газов алмазодобывающих рудников западной Якутии. *Маркшейдерия и Недропользование* 49: 30–34.
71. Гумен, А. and А. Гусев, Газогеохимические индикаторы геодинамической активности глубинных разломов на юго-востоке Беларуси. Літасфера 1997: p. 140–149.
72. Гусев, А., Газогеохимические эффекты современной геодинамической активности платформенных структур (на примере Юго-Восточной Белоруссии). 1997.
73. Ballentine, C. J., R. K. O'Nions, and M. L. Coleman. 1996. A Magnus opus: Helium, neon, and argon isotopes in a North Sea oilfield. *Geochimica et Cosmochimica Acta* 60 (5): 831–849.
74. Fallick, A. E., C. I. Macaulay, and R. S. Haszeldine. 1993. Implications of linearly correlated oxygen and hydrogen isotopic compositions for kaolinite and illite in the Magnus Sandstone. *North Sea. Clays and Clay Minerals* 41 (2): 184–190.
75. Finlay, A., M. Osborne, and D. Finucane. 2010. Fault-charged mantle-fluid contamination of United Kingdom North Sea oils: Insights from Re-Os isotopes. *Geology* 38: 979.
76. Нечаева, О. 1968. К вопросу о водороде в газах, растворенных в водах Западно-Сибирской низменности. *Доклады АН СССР* 179: 961–962.
77. Ostojic, S., Are there natural spring waters rich in molecular hydrogen? Trends in Food Science & Technology, 2019. 90.
78. Lin, L.H., et al., Radiolytic H_2 in continental crust: Nuclear power for deep subsurface microbial communities. Geochemistry Geophysics Geosystems, 2005. 6.
79. Sidle, W.C. and J.V.T.L. Hydrocarbon, Hydrogen and helium anomalies in the eastern Snake River, Idaho. In: 183rd ACS Natl. Meet., Las Vegas, NV, 1982.
80. Овчаренко, Ю., Водород в газах равнинного Крыма. Геология и Геохимия Горючих Ископаемых Геология и Нефтегазоносность Причерноморской Впадины, 1967: p. 90–97.
81. Зингер, А. 1962. Молекулярный водород в составе газа, растворенного в водах газонефтяных месторождений Нижнего Поволжья. *Геохимия* 10: 890–898.
82. Bjerg, P. L., et al. 1997. Effects of sampling well construction on H_2 measurements made for characterization of redox conditions in a contaminated aquifer. *Environmental Science & Technology* 31 (10): 3029–3031.
83. Chapelle, F. H., et al. 1996. Comparison of Eh and H_2 measurements for delineating redox processes in a contaminated aquifer. *Environmental Science & Technology* 30 (12): 3565–3569.
84. Bourdet, J., et al., Natural hydrogen in low temperature geofluids in a Precambrian granite, South Australia. Implications for hydrogen generation and movement in the upper crust. Chemical Geology, 2023. 638.
85. Giuliani, A., et al. 2023. Genesis and evolution of kimberlites. *Nature Reviews Earth & Environment* 4 (11): 738–753.
86. Wang, L., et al. 2023. The origin and occurrence of natural hydrogen. *Energies* 16 (5): 2400–2400.
87. Niu, Y., et al. 2024. Exploring the potential of microbial coalbed methane for sustainable energy development. *Molecules* 29 (15).

7 Caprock and Sealing Integrity

H_2, being a small and highly mobile molecule, can easily escape through even tiny cracks or faults in rock formations. Therefore, an effective seal is crucial for trapping H_2. The most effective caprocks are those with very low permeability, such as salt, anhydrite, and tightly compacted shale, which prevent H_2 from diffusing out of the reservoir [1].

The overlying caprock works effectively due to the extremely small molecular size, high diffusivity, and low density of H_2, it exhibits enhanced permeability through geological formations and presents challenges for effective containment by conventional caprocks. Compared to conventional oil and gas caprocks such as mudstone and shale, natural hydrogen system requires caprocks with superior compactness and lower permeability [2]. Therefore, caprock lithologies are significant, lithologies such as mudstone, shale, halite, and tight carbonate rocks, incorporating a comprehensive analysis of hydrogen's molecular dimensions, wettability characteristics, and capillary pressure dynamics.

Compared to hydrocarbon gases, H_2 exhibits more active physicochemical properties and is more prone to migration and dissipation. Therefore, the sealing capacity of caprocks constitutes a critical controlling factor for natural hydrogen accumulation, with hydrogen reservoirs requiring more stringent physical properties and mechanical strength characteristics in caprocks. Case studies reveal that caprock characteristics of natural hydrogen system share similarities but also exhibit distinct differences compared to petroleum systems. For instance, Paleozoic mudstone-shale formations serve as primary caprocks for natural hydrogen system in Kansas, USA, and Kangaroo Island, Australia [3], while salt formations cap the H_2 reservoirs in the northern Pyrenees. In Mali, the caprocks predominantly consist of later intruded diabase sills [4, 5], where a positive correlation has been observed between sill thickness and H_2 concentration, indicating caprock thickness

Y. Yuan et al., *Natural Hydrogen*, Synthesis Lectures on Renewable Energy Technologies, https://doi.org/10.1007/978-3-032-14469-0_7

as a crucial parameter influencing H_2 sealing efficiency. The reservoir-seal configuration of the H_2 accumulation in the Bulqizë chromite deposit, Albania, presents unique characteristics. Truche et al. propose that this natural hydrogen accumulation is primarily concentrated within ultramafic rock masses underlying the chromite ore [6]. Fracture zones within the ultramafic mass provide favorable conditions for H_2 enrichment, while the overlying dense rock mass with limited fracture development effectively seals the H_2 reservoir (Fig. 7.1). The discovery of this accumulation resulted from mining activities that penetrated the tight caprock, inducing H_2 leakage during chromite extraction [7].

The textural and compositional characteristics of sedimentary rocks, including porosity, grain size distribution, and degree of cementation, govern H_2 diffusion dynamics and reservoir potential within layered stratigraphic sequences. Sedimentary permeability gradients, controlled by lithological variations, dictate H_2 migration pathways and entrapment efficiency beneath low-permeability seal units. Concordant magmatic intrusions (e.g., sill complexes) function as transient geological barriers within sedimentary basins, temporarily impounding H_2 through permeability contrasts. This mechanism facilitates localized H_2 enrichment, exemplified by dolerite-hosted accumulations documented in the Jurassic strata of the Lonnavale 1 well. Furthermore, compressional tectonism generating folded structures (e.g., anticlinal closures and synclinal troughs) establishes three-dimensional traps for H_2 concentration, particularly when structural geometry combines with overlying impermeable units to form effective hydrodynamic seals [8–10].

Intrusions represent a critical factor controlling the distribution and accumulation of H_2 within the lithosphere. Their impact is particularly profound in channeling and temporarily trapping H_2. For instance, concordant intrusions function as natural caprocks, sealing transient H_2 reservoirs and inhibiting their upward migration. This sealing mechanism is exemplified by the Bougou-1 well (located in Mali) and the Lonnavale 1 well (located in Tasmania, Australia), where Jurassic dolerite intrusions function as effective caprocks for underlying hydrogen accumulations [11]. In the Lonnavale 1 well, H_2 concentrations of up to 85% were reported, and the dolerite intrusion covering approximately three-quarters of the basin, thereby forming the principal regional seal for the underlying sedimentary succession [12]. Notably, fault systems are also likely contributors to regulating H_2 migration in these settings, underscoring the interplay between tectonic structures and intrusive bodies in governing H_2 dynamics [9].

The investigation reveals that salt caverns present the minimal hydrogen leakage risk among the three subsurface storage options: salt caverns, depleted oil and gas reservoirs, and aquifers. The dense nature of salt rock provides effective sealing capacity for H_2 containment. However, creep deformation of salt rock induced by variations in forma-

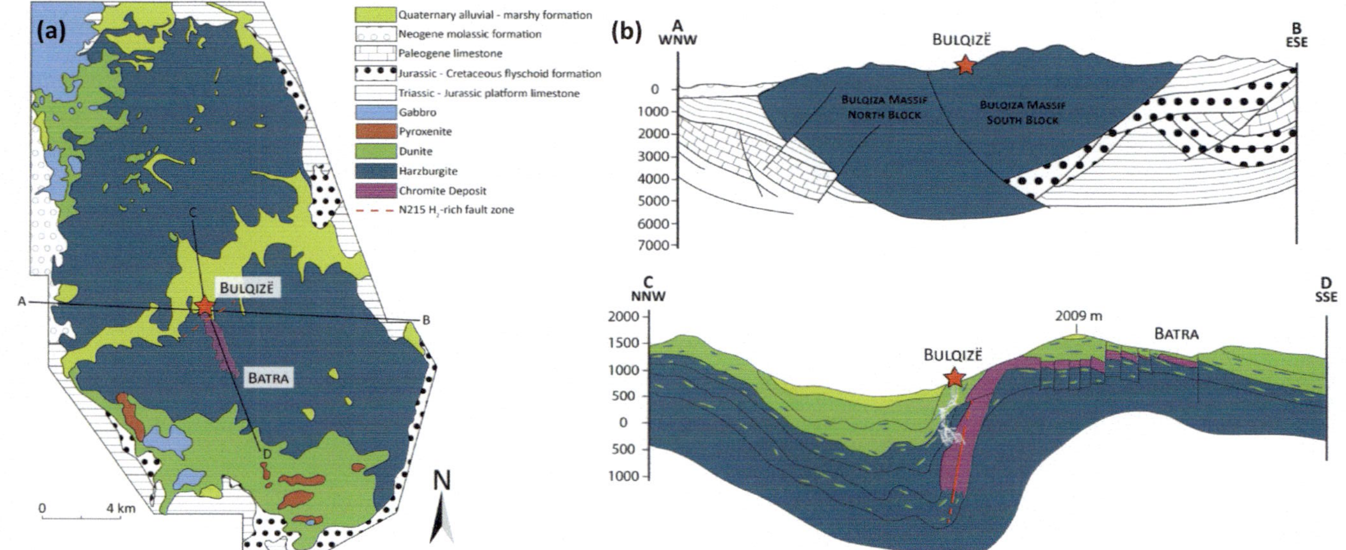

Fig. 7.1 Three-dimensional map of the Bulqizë ultramafic massif and its controlling fault zone. **a** Generalized geologic map of the Bulqizë ultramafic massif, illustrating the surface expression of the ore deposit and the deep-seated, H_2-rich fault zone oriented N215° with an 80° W dip. **b** Geological cross-sections along the A-B and C-D transects (reproduced with permission from [6])

tion temperature and pressure conditions may generate fractures on mineral surfaces, thereby compromising its H_2 sealing efficiency [13]. The sealing capacity of caprocks primarily depends on pore size and tortuosity, where smaller pore dimensions (< 5 nm) and higher tortuosity (> 3) significantly enhance containment effectiveness [14–18]. Beyond pore geometry, H_2 diffusivity shows strong dependence on subsurface environmental conditions. Elevated reservoir temperatures accelerate leakage rates, although elevated pressure decreases the mean free path of H_2 molecules, it suppresses diffusional mobility and improves hydrogen retention. When evaluating caprock integrity, the high reactivity of H_2 necessitates comprehensive consideration of multiple factors, including H_2 fluid properties, PVT characteristics of H_2-reservoir fluid (oil–gas-water) systems, geochemical reactions, microbial activities, and H_2-induced alterations in caprock petrophysical/mechanical properties [19]. Under subsurface pressure–temperature conditions, H_2 interacts with formation fluids and rocks can significantly modify caprock mechanical behavior, affecting its brittle-ductile transition and potentially inducing fracture propagation that facilitates H_2 escape [20, 21]. Therefore, caprock condition assessment for natural hydrogen system constitutes a complex process requiring multivariate analysis. This evaluation should incorporate not only the inherent sealing capacity of caprocks but also mechanical property alterations induced by hydrogen's physicochemical reactivity. The comprehensive understanding of these interdependent factors is crucial for ensuring the long-term integrity and safety of underground H_2 storage systems [7].

As illustrated in Fig. 7.2, H_2-brine-rock interactions may induce redox reactions, mineral dissolution/precipitation, pH variations, ionic strength alterations, and fracture reactivation/propagation, thereby potentially compromising the integrity of caprock formations. Furthermore, subsurface-stored H_2 may react with certain minerals or microbial communities, leading to hydrogen alteration and potential degradation of subsurface hydrogen reservoirs [22]. It is evident that unconsolidated lithologies such as uncemented sandstones, loose marls, and highly fractured schists exhibit higher permeability to hydrogen compared to intrusive bodies, particularly concordant intrusions like sills. In some cases, the latter factor plays a decisive role in the development of a seal, thereby establishing a temporary "trap" for hydrogen accumulation, a behavior that exhibits certain parallels with that of hydrocarbon gases.

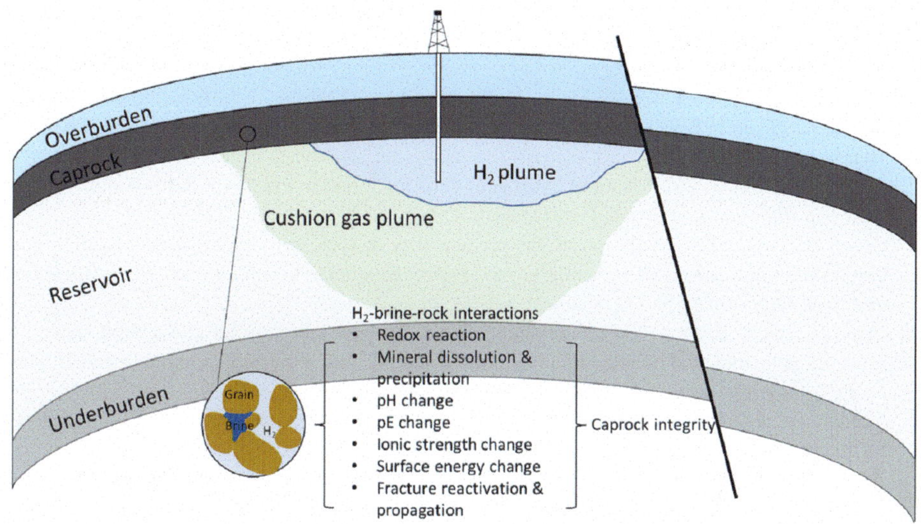

Fig. 7.2 Schematic representation of the impact of H_2-brine-rock interactions on caprock sealing integrity during underground hydrogen storage (reproduced with permission from [23])

References

1. Babatunde, K., and H. Emami-Meybodi. 2025. Gas transport in shales with applications to geological storage of H_2 and CO_2. *Gas Science and Engineering* 135: 205550.
2. Alafnan, S. 2024. Assessing leakage risks of hydrogen through aquifers and caprocks. *Energy & Fuels* 38 (20): 19739–19747.
3. Newell, K. D., et al. 2007. H_2-rich and hydrocarbon gas recovered in a deep Precambrian well in northeastern Kansas. *Natural Resources Research* 16 (3): 277–292.
4. Hutchinson, I.P., et al. 2024. Greenstones as a source of hydrogen in cratonic sedimentary basins. *Geological Society* 547 (1): SP547-2023-39.
5. Maiga, O., et al. 2024. Trapping processes of large volumes of natural hydrogen in the subsurface: The emblematic case of the Bourakebougou H_2 field in Mali. *International Journal of Hydrogen Energy* 50: 640–647.
6. Truche, L., et al. 2025. A dynamic H2 system with multi-source methane in chromitite-rich ophiolitic settings. Geochimica et Cosmochimica Acta 409: 281-307
7. Yin, L., et al. 2024. Origins and accumulation characteristics of large-scale generation of natural hydrogen (in Chinese). *Lithologic Reservoirs* 36 (06): 1–11.
8. Wei, Q., et al. 2023. Geological characteristics, formation distribution and resource prospects of natural hydrogen reservoir (in Chinese). *Natural Gas Geoscience* 35 (6): 1113–1122.
9. Vitaly, V., and R. Reza. 2022. Natural deep-seated hydrogen resources exploration and development: Structural features, governing factors, and controls. *Journal of Energy and Natural Resources* 11 (3): 60–81.

10. Tian, Q., et al. 2022. Origin, discovery, exploration and development status and prospect of global natural hydrogen under the background of "carbon neutrality. *China Geology* 5 (4): 722–733.
11. Prinzhofer, A., C. S. T. Cissé, and A. B. Diallo. 2018. Discovery of a large accumulation of natural hydrogen in Bourakebougou (Mali). *International Journal of Hydrogen Energy* 43 (42): 19315–19326.
12. Emanuelle, F., et al. 2021. Natural hydrogen seeps identified in the North Perth Basin, Western Australia. *International Journal of Hydrogen Energy* 46 (61): 31158–31173.
13. Grgic, D., et al. 2021. Evolution of gas permeability of rock salt under different loading conditions and implications on the underground hydrogen storage in salt caverns. *Rock Mechanics and Rock Engineering* 55 (2): 1–24.
14. Yuan, Y., and R. Rezaee. 2019. Fractal analysis of the pore structure for clay bound water and potential gas storage in shales based on NMR and N2 gas adsorption. *Journal of Petroleum Science and Engineering* 177: 756–765.
15. Yuan, Y., et al. 2019. Impact of composition on pore structure properties in shale: Implications for micro-/mesopore volume and surface area prediction. *Energy & Fuels* 33: 9619–9628.
16. Alafnan, S. 2024. Factors influencing hydrogen migration in cap rocks: Establishing new screening criteria for the selection of underground hydrogen storage locations. *International Journal of Hydrogen Energy* 83: 1099–1106.
17. Liu, K., et al. 2019. A comprehensive pore structure study of the Bakken Shale with SANS, N2 adsorption and mercury intrusion. Fuel 245: 274–285.
18. Yuan, Y., et al. 2021. Compositional controls on nanopore structure of different shale lithofacies: A comparison with pure clays and isolated kerogens. Fuel 303: 121079.
19. Li, F., et al. 2023. Underground hydrogen storage technology: storage mechanism, leakage risk and existing problems (in Chinese). In *2023 International Field Exploration and Development Conference*, Wuhan, China.
20. Dilshan, R.A.D.P., M.S.A. Perera, and S.K. Matthai. 2024. Effect of mechanical weakening and crack formation on caprock integrity during underground hydrogen storage in depleted gas reservoirs—A comprehensive review. Fuel 371 (PA): 131893.
21. Yuan, Y., et al. 2023. A comprehensive review on shale studies with emphasis on nuclear magnetic resonance (NMR) technique. *Gas Science and Engineering* 120: 205163.
22. Wang, L., et al. 2024. Research progress in underground hydrogen storage (in Chinese). *Earth Science* 49 (06): 2044–2057.
23. Zeng, L., et al. 2023. Role of geochemical reactions on caprock integrity during underground hydrogen storage. *Journal of Energy Storage* 65.

Trap and Retention of Natural Hydrogen

8

For the migrated H_2 molecules to be accumulated in economically viable quantities, trap within suitable geological formations is generally required. Similar to the mechanisms that trap hydrocarbons, H_2 can be stored in structural traps, stratigraphic traps, or combination traps (Fig. 8.1) [1– 4]. Traditional traps can be classified into three main types. (i) Structural Traps: These traps are formed by the deformation of rock layers due to tectonic forces, creating structures such as anticlines, fault-bounded compartments, or folded rock layers 5–7. H_2 can accumulate beneath impermeable layers (caprocks), such as shale or salt, that prevent it from escaping. Faults and fold structures frequently serve as critical seals that confine and stabilize natural hydrogen accumulations. (ii) Stratigraphic Traps: Stratigraphic traps form when changes in rock type (such as from porous sandstone to impermeable shale) create a natural barrier to H_2 migration. In these settings, H_2 migrates through permeable rocks but becomes trapped when it encounters impermeable layers that prevent further upward movement. (iii) Combination Traps: A mix of structural and stratigraphic elements, where rock deformation and variations in rock types work together to create a sealed environment for H_2 accumulation.

Salt formations, such as those in Queensland's Adavale Basin, serve as effective reservoirs for H_2 storage. The inherent plasticity and low permeability of salt enable it to self-seal fractures, effectively preventing H_2 migration over time [9]. Salt caverns are regarded as the premier option for large-scale underground hydrogen storage owing to their distinctive geomechanical characteristics. These geological structures not only ensure commercial viability but also provide critical safety assurances required for long-term containment.

In summary, the hydrogen gas accumulated in these reservoirs predominantly originates from crustal-scale chemical processes, notably serpentinization (a reaction involving

Y. Yuan et al., *Natural Hydrogen*, Synthesis Lectures on Renewable Energy Technologies, https://doi.org/10.1007/978-3-032-14469-0_8

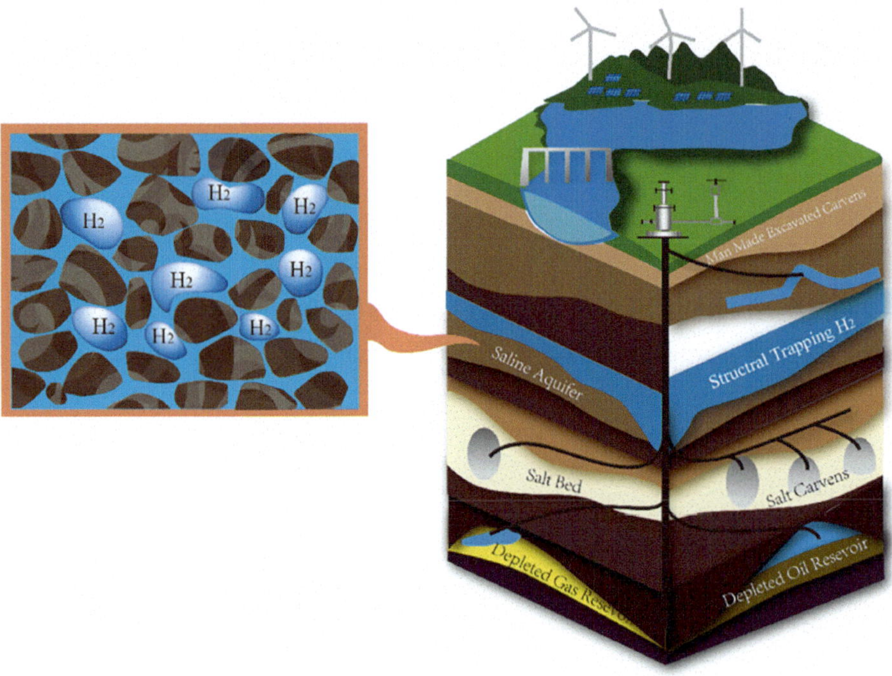

Fig. 8.1 Structural trapping, capillary trapping, and mineral trapping are depicted in the aquifer-based reservoir system (right). Hydrogen molecules migrate upward and may become trapped through residual and structural mechanisms (left) (adapted from [1, 8])

water and ultramafic lithologies such as peridotite) and radiolysis (the dissociation of water occurs under radioactive conditions) [10]. Over time, these processes lead to the accumulation of H_2, which is trapped by impermeable rock layers, much like oil and gas [11]. However, compared to Hydrocarbon Traps, traditional hydrocarbon traps often coexist with other gases, but pure H_2 traps are increasingly being identified, indicating the possibility of extracting clean H_2 without CO_2 or other harmful emissions. Due to its small molecular size and high diffusion coefficient, hydrogen necessitates the presence of highly effective caprocks and impermeable seals to minimize leakage [12].

The entrapment of natural hydrogen system represents a complex and crucial concept in geology, referring to subsurface geological settings where H_2 is effectively accumulated and preserved through geological processes [13]. A caprock denotes the overlying geological stratum above a reservoir that prevents upward migration of fluids such as hydrocarbons or H_2. It generally exhibits low permeability and elevated capillary pressure, enabling effective confinement of subsurface fluids. In natural hydrogen accumulations, the presence of a competent caprock constitutes one of the essential prerequisites for the formation of commercially viable reservoirs [11]. Based on geological characteristics and genetic mechanisms, caprocks can be classified into multiple distinct types.

8.1 Structural Trap

Structural traps form through tectonic deformation (faults, folds) that generate impermeable barriers, thereby inhibiting the migration of H_2. Faults and Folds are shown in Fig. 8.2. These geological structures serve as critical pathways for H_2 migration, functioning as either conduits or barriers depending on their geometry and subsurface context. Fault zones may significantly increase subsurface permeability, facilitating H_2 migration through fractured rock networks, whereas folds can accumulate H_2 by forming structural traps, particularly within anticlinal closures or other elevated structural configurations [14].

Structural traps refer to sealed layers formed by geological tectonic processes (such as faults and folds). These tectonic activities alter the original structure and arrangement of rocks, creating barriers to fluid migration. (i) Fault sealing: In certain cases, faults can act as caprocks for natural hydrogen system. When there are significant differences in lithological properties between the two walls of a fault, the argillaceous fillings or brecciated rocks within the fault zone may exhibit low permeability, thereby preventing hydrogen escape. However, it should be noted that not all faults can form effective seals. The openness, activity, and filling properties of faults will influence their sealing capacity [16]. (ii) Fold limb sealing: In fold structures, limb rocks may become compacted and exhibit reduced permeability due to compressive forces, thereby forming seals. This type of seal is commonly observed in areas with well-developed fold structures [17].

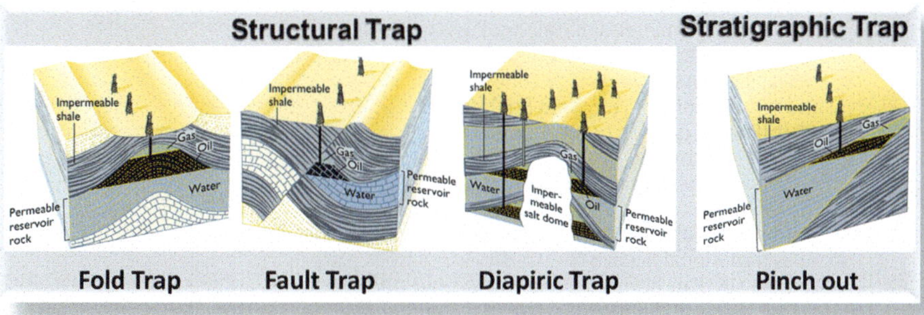

Fig. 8.2 The traditional trap types, including structural trap (e.g., fold trap, fault trap, and diapiric trap) and stratigraphic trap (e.g., pinch out) (adapted from [15])

8.2 Stratigraphic Trap

Stratigraphic Traps are formed due to changes in rock characteristics such as porosity and permeability. Lithologic caprocks are primarily composed of low-permeability rocks. Sedimentary formations such as mudstone, shale, and evaporites (e.g., gypsum, halite) exhibit dense microstructures and low porosity, effectively preventing H_2 escape through pores and fractures [11]. (i) Mudstone caprocks: As a common lithological seal, mudstone serves as an effective containment layer for natural gas and H_2 reservoirs due to its low permeability and excellent ductility. For instance, mudstone layers in certain Precambrian iron-rich formations may act as natural hydrogen seals. (ii) Evaporite caprocks: Evaporitic minerals like gypsum and halite constitute premium sealing materials owing to their extremely low permeability, resulting from high-density mineral crystal formation during sedimentation. In specific rift settings, evaporite caprocks have demonstrated effective confinement of natural hydrogen accumulations.

8.3 Hydrodynamic Trapping

In groundwater systems, hydraulic flow is governed by multiple factors, including topography, lithology, and structural features, which collectively generate intricate flow networks [18]. When these water flows encounter barriers of low-permeability strata or structural fractures within specific geological environments, stagnant flow zones or retention areas develop in the obstructed regions. Within these stagnant zones, significant alterations occur in hydrodynamic conditions, where flow velocities diminish or even stagnate, resulting in the gradual exsolution and accumulation of dissolved gases (e.g., H_2) from the aqueous phase. Due to persistent variations in hydrodynamic conditions, these accumulated gases are often preserved over extended periods within such geological settings [15–21].

8.4 Examples

8.4.1 Pyrenees

Natural hydrogen occurrences have been reported from both the French and Spanish sides of the suture zone [26, 27]. As a result of the subduction and collision between the Iberian Plate and Eurasian Plate, mantle-derived materials upwelled and formed Fe–Mg-rich peridotites in shallow crustal levels, where subsequent serpentinization generated H_2 that migrates along fault zones into sedimentary basins (Fig. 8.3). The caprock of natural hydrogen system on the northern flank of the Pyrenees consists of salt rock.

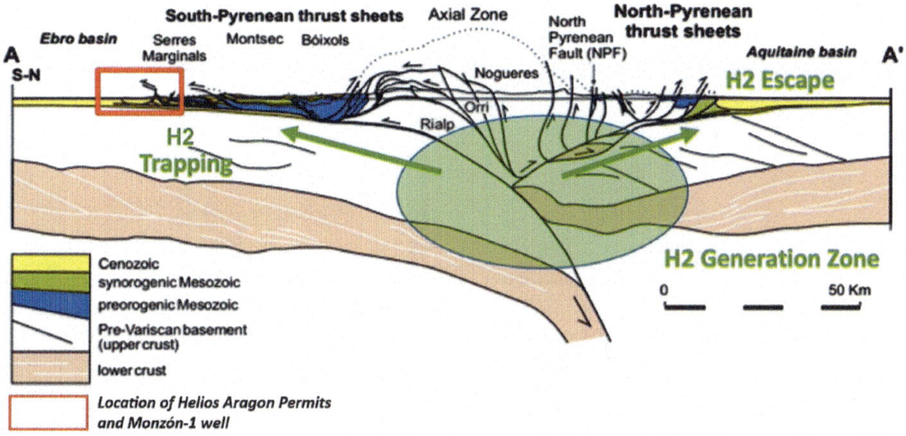

Fig. 8.3 Natural hydrogen accumulation model in the Pyrenees (reproduced with permission from [28])

8.4.2 Taoudeni Basin, Mali

In the Precambrian strata of the Bourakebougou area, banded iron formations (BIFs) have developed, serving as one of the key H_2 sources for this natural hydrogen system [29]. The redox reaction between Fe^{2+} in the iron formations and water generates H_2, which migrates through pathways such as fault zones and accumulates in overlying reservoirs to form accumulations (Fig. 8.4). Therefore, regions enriched with Precambrian banded iron formations (BIFs) represent favorable targets for natural hydrogen exploration [30]. The caprock of Mali's natural hydrogen system primarily consists of later-stage sill-shaped dolerite intrusions, with a distinct positive correlation observed between dolerite thickness and H_2 concentration, demonstrating that caprock thickness constitutes a critical parameter influencing H_2 sealing capacity.

8.4.3 Kansas Basin

The development of rift systems is closely associated with plate tectonic movements, and is driven by mantle-derived upwelling in conjunction with extensional deformation of the crust. Consequently, these systems are often characterized by long-lived deep-rooted fault systems and widespread intense mafic–ultramafic magmatism, which together provide conducive conditions for the generation and migration of natural hydrogen. [10]. Natural hydrogen accumulations in such geological settings typically exhibit multiple genetic origins, including water–rock reactions in iron-rich lithologies and mantle degassing, making these regions promising targets for large-scale H_2 exploration. The Kansas Basin is

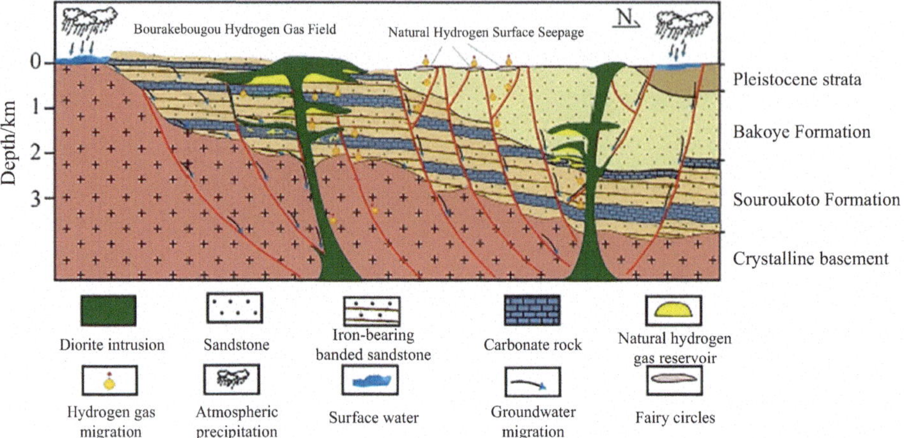

Fig. 8.4 Geological model of natural hydrogen system in the Bourakebougou region, Taoudeni Basin, Mali (adapted from [29])

situated along the western flank of the rift system, and high concentrations of natural hydrogen have been detected in over a dozen exploration wells [31]. All of these wells are located along the Nemaha Anticline, to the west of the Humboldt Fault within the rift zone (Fig. 8.5). Long-term monitoring of H_2 concentrations in this area demonstrates continuous H_2 generation (Fig. 8.6). Through research on the causes of the region, it is believed that H_2 in the region is not only a source of mantle degassing, but also derives from water–rock reactions involving Fe^{2+}bearing minerals in the Mississippian and Pennsylvanian Systems. The resulting H_2 migrates through groundwater systems to accumulate in reservoirs [29]. The caprocks for natural hydrogen accumulations in central Kansas (United States) primarily consist of Paleozoic shale formations.

8.5 Preserve

H_2 is a highly reactive molecule capable of combining with nearly all elements (the exception of noble gases). As a primary constituent of water molecules, various minerals, organisms, organic matter, and hydrocarbons, H_2 participates in multiple consumption processes within the Earth's lithosphere. These include H_2-mineral/rock reactions, microbial H_2 consumption, and hydrocarbon generation through hydrogenation. Understanding the mechanisms of these subsurface H_2-consuming processes is crucial for investigating the accumulation and preservation of natural hydrogen resource. The predominant H_2 consumption mechanisms in subsurface environments primarily involve: microbial metabolic utilization of hydrogen, and hydrogenation-induced hydrocarbon generation in mid-deep formations. In anaerobic subsurface environments, certain microorganisms

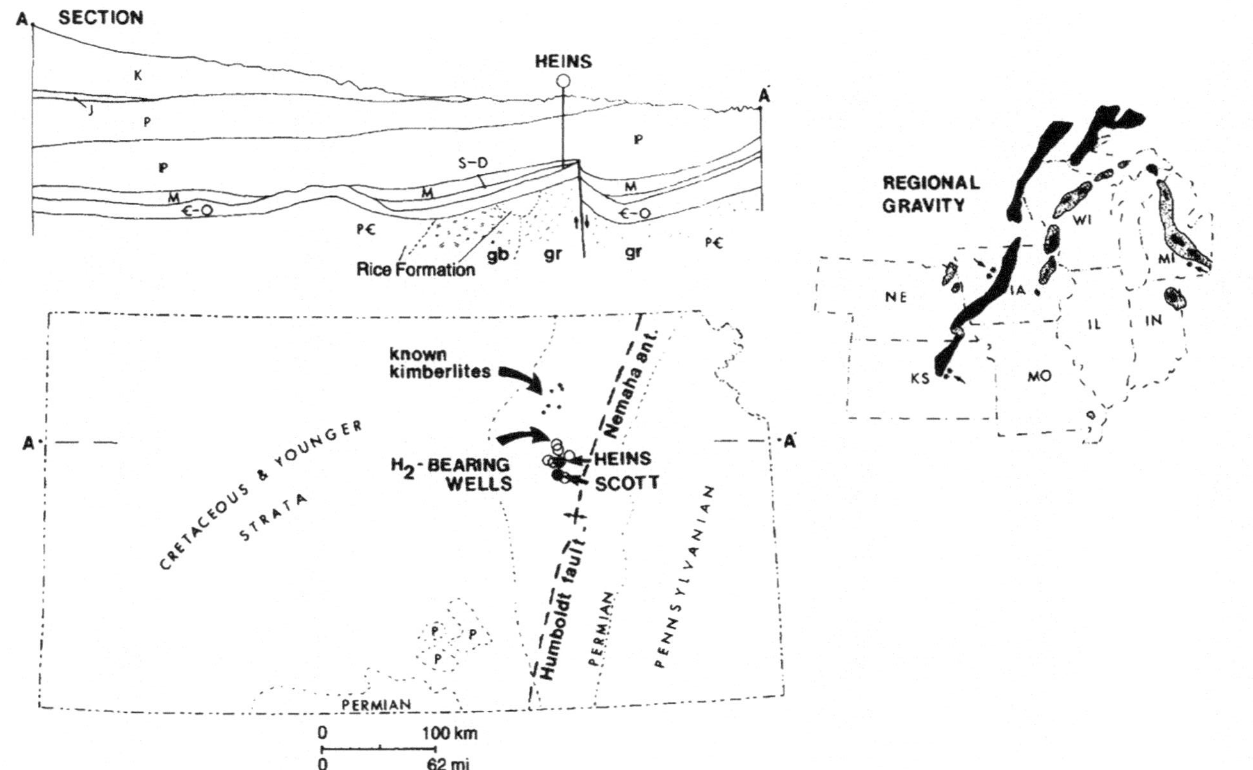

Fig. 8.5 Geological model of Kansas and regional gravity characteristics of the northern Midwestern United States. The regional gravity map and cross-section AA′ are adapted from Bickford et al. (1979) and the Kansas Geological Survey (1984). The geologic cross-section AA′ (schematic) roughly traces the route of Interstate 70 across Kansas (adapted from [32])

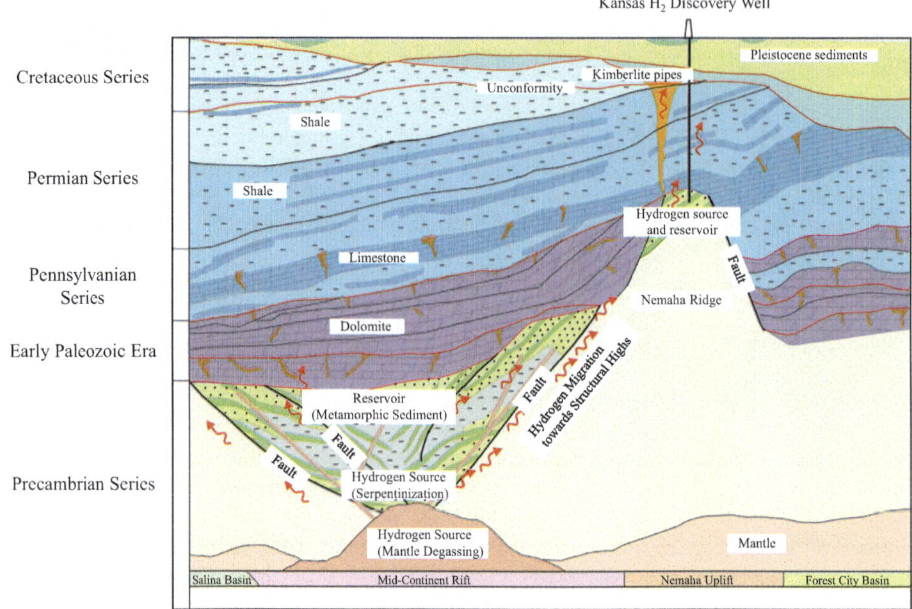

Fig. 8.6 Geological model of natural hydrogen system and spillage from the Kansas segment of the Midcontinental Rift System in the United States (adapted from [33, 34])

generate substantial H_2 during organic matter degradation, while others (termed hydrogen-consuming microorganisms [35], such as hydrogenotrophic methanogens [36], utilize this H_2 for their metabolic processes. During anaerobic fermentation, hydrogenotrophic methanogens consume CO_2 and H_2 as substrates to produce CH_4 through biochemical reactions ($4H_2 + CO_2 \rightarrow CH_4 + 2H_2O$) [37]. Such microbial activities are widespread in geological strata, with approximately 70% of atmospheric methane emissions originating from anaerobic microbial methanogenesis [38]. When sufficient H_2 sources, substantial methane production, and favorable preservation conditions coexist, these processes can form biogenic gas reservoirs - a significant natural gas resource [29].

In fact, microbial gas is widely distributed, accounting for approximately 15%–20% of the total proven natural gas reserves [39]. The Sebei Gas Field in the eastern Qaidam Basin, China, represents such a microbial gas reservoir. Studies have revealed well-developed bacterial profiles within the biogas-producing layers of the Sebei Formation [39]. Experimental investigations using Sebei core samples demonstrated accelerated methane production rates and enhanced yields following CO_2 and H_2 injection [40]. Notably, Shuai Yanhua et al. detected high H_2 concentrations in this region. This suggests that substantial H_2 generation occurs in the Sebei strata, which is subsequently utilized by methanogens as substrate for large-scale CH_4 conversion, ultimately forming commercial microbial gas accumulations [41, 42]. H_2 serves as both a critical intermediate and primary product during petroleum generation. During the thermal maturation of organic matter into petroleum, numerous intermediate compounds are generated that are inherently hydrogen-deficient, necessitating H_2 supplementation for processes including the decomposition of macromolecular compounds into smaller molecules and the saturation of olefins into paraffinic hydrocarbons [35].

In general, H_2 transfer within these organic compounds maintains a stable dynamic equilibrium. However, when exogenous H_2 is introduced, hydrocarbon evolution undergoes accelerated conversion from heavy hydrocarbons to light methane during the stages spanning from catagenesis to metagenesis, accompanied by increased yields of light oil and methane [43]. Remarkably, even at the semi-graphitization stage of organic matter, Fischer–Tropsch synthesis can occur to generate CH_4 through the addition of external H_2. Scholars investigating deep natural gas reservoirs in the Tarim and Sichuan Basins have proposed an organic–inorganic composite hydrocarbon generation mechanism, demonstrating the involvement of deep-sourced inorganic H_2 in hydrocarbon formation [29].

In summary, subsurface H_2 consumption makes H_2 accumulation and detection challenging in sedimentary environments. However, increasing exploration practices have revealed the existence of naturally occurring H_2 accumulations at scale. These exploration cases suggest that natural hydrogen exploration should follow the principle of H_2 input exceeding consumption, focusing on favorable generation areas such as ophiolite development zones, Precambrian banded iron formation (BIF) belts, and rift systems. Concurrently, regions with substantial H_2 consumption mechanisms should be avoided, including areas of microbial H_2 consumption and hydrocarbon reservoirs undergoing hydrogenation. Consequently, it is proposed that ultra-deep strata, peripheral zones of major hydrocarbon accumulation areas, and shallow strata near deep-seated faults represent potential favorable domains for natural hydrogen accumulation (Fig. 8.7) [24].

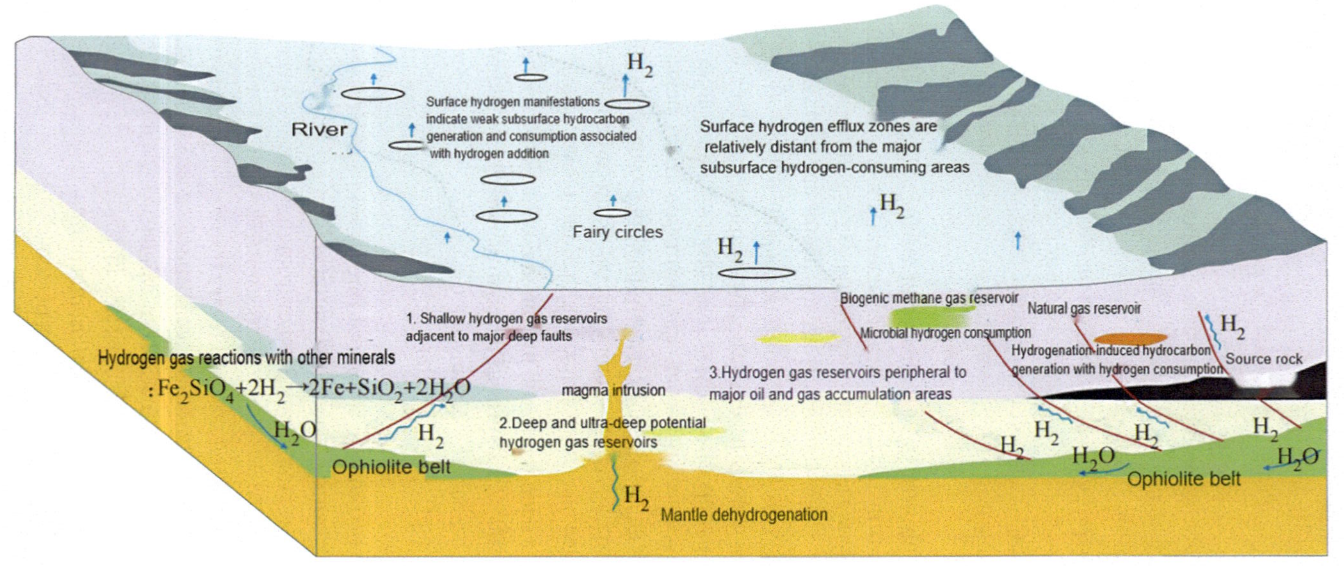

Fig. 8.7 Schematic representation of prospective zones for natural hydrogen exploration (adapted from [26])

References

1. Liu, J. Y., et al. 2023. Genesis and energy significance of natural hydrogen. *Unconventional Resources* 3: 176–182.
2. Prinzhofer, A., et al. 2019. Natural hydrogen continuous emission from sedimentary basins: The example of a Brazilian H_2-emitting structure. *International Journal of Hydrogen Energy* 44 (12): 5676–5685.
3. Tian, Q., et al. 2022. Origin, discovery, exploration and development status and prospect of global natural hydrogen under the background of "carbon neutrality." *China Geology* 5 (4): 722–733.
4. Epelle, E. I., et al. 2022. Perspectives and prospects of underground hydrogen storage and natural hydrogen. *Sustainable Energy & Fuels* 6 (14): 3324–3343.
5. Al-Khdheeawi, E. A., et al. 2024, Production logging tools interpretation for a vertical oil well. *Petroleum Chemistry* 1–9.
6. Al-Khdheeawi, E.A., et al. 2024. Well spacing optimization to enhance the performance of tight reservoirs. *Petroleum Chemistry* 64: 829–839
7. Mahdi, D.S., et al. 2025. AI-Based Estimation of Poisson's Ratio for Carbonate Formations Using Drilling Parameters in a Southern Iraqi Oil Field. *Journal of Petroleum Research and Studies* 15 (2): 44–59.
8. Aftab, A., et al. 2022. Toward a fundamental understanding of geological hydrogen storage. *Industrial & Engineering Chemistry Research* 61 (9): 3233–3253.
9. Bradshaw, J. D. 2023. The Ross-Delamerian Orogen in the southwest Pacific and Antarctica: An active plate boundary for Gondwana in the late Neoproterozoic and Cambrian. *New Zealand Journal of Geology and Geophysics* 66 (3): 374–397.
10. Zgonnik, V. 2020. The occurrence and geoscience of natural hydrogen: A comprehensive review. *Earth-Science Reviews* 203: 103140.
11. Nagumo, M. 2023. *Hydrogen Trapping and Its Direct Detection*, 11–40.
12. Panfilov, M., G. Gravier, and S. Fillacier. 2006. Underground storage of H_2 and H_2-CO_2-CH_4 mixtures. In *ECMOR 2006—10th European Conference on the Mathematics of Oil Recovery*.
13. Ballentine, C. J., et al. 2025. Natural hydrogen resource accumulation in the continental crust. *Nature Reviews Earth & Environment* 6 (5): 342–356.
14. Aydin, A. 2000. Fractures, faults, and hydrocarbon entrapment, migration and flow. *Marine and Petroleum Geology* 17 (7): 797–814.
15. Presentation slide of Peter Copeland and William Dupre at Houston University. Open-source link: https://www.slideserve.com/sade-jordan/classroom-presentations-to-accompany-understanding-earth-3rd-edition
16. Zhang, J. 2021. Research on fault sealing mechanism and hydrocarbon migration and accumulation characteristics (in Chinese).
17. Eckert, A., P. Connolly, and X. L. Liu. 2014. Large-scale mechanical buckle fold development and the initiation of tensile fractures. *Geochemistry Geophysics Geosystems* 15 (11): 4570–4587.
18. Lou, Z., et al. 2011. Hydrocarbon preservation conditions in marine strata of the Guizhong Depression and its margin (in Chinese). *Shiyou Xuebao/Acta Petrolei Sinica* 32: 432–441.
19. Shuai, Y. H., et al. 2013. Controls on biogenic gas formation in the Qaidam Basin, northwestern China. *Chemical Geology* 335: 36–47.
20. Malachowska, A., et al. 2022. Hydrogen storage in geological formations—The potential of salt caverns. Energies 15(14).
21. Wang, L., et al. 2023. The origin and occurrence of natural hydrogen. *Energies* 16 (5): 2400–2400.

22. Boruah, A., P. Alaktaraag, and S. Singh, Utilization of coal for hydrogen generation. *International Journal of Coal Preparation and Utilization* 1–22.
23. Sazali, N., M. A. Mohamed, and W. N. W. Salleh. 2020. Membranes for hydrogen separation: A significant review. *International Journal of Advanced Manufacturing Technology* 107 (3–4): 1859–1881.
24. Du, Z.M., et al., A review of hydrogen purification technologies for fuel cell vehicles. *Catalysts* 11(3).
25. Coyle, S. 2022. Hydrogen storage potential of the Salina Group, Appalachian and Michigan basins.
26. Paul, H., et al. 2021. Hydrogen gas in circular depressions in South Gironde, France: Flux, stock, or artefact? *Applied Geochemistry* 127 (prepublish): 104928.
27. Atkinson, C., et al. 2023. Geological setting of natural "gold" hydrogen in the Pyrenees and implications for exploration worldwide.
28. Natural hydrogen in the Monzón-1 well, Ebro basin, northern Spain-Geological Society of France publication-Helios [EB/OL]. (2022-06-20)[2024-08-30]. https://helios-aragon.com/naturalhydrogen-in-the-monzon-1-well-ebro-basin-northern-spain-geological-society-of-france-publication/.
29. Yin, L., et al. 2024. Origins and accumulation characteristics of large-scale generation of natural hydrogen (in Chinese). *Lithologic Reservoirs* 36 (06): 1–11.
30. Prinzhofer, A., C. S. T. Cissé, and A. B. Diallo. 2018. Discovery of a large accumulation of natural hydrogen in Bourakebougou (Mali). *International Journal of Hydrogen Energy* 43 (42): 19315–19326.
31. Guélard, J., et al. 2017. Natural H_2 in Kansas: Deep or shallow origin? *Geochemistry, Geophysics, Geosystems* 18 (5): 1841–1865.
32. Coveney, R. M., et al. 1987. Serpentinization and the origin of hydrogen gas in Kansas. *GeoScienceWorld* 71 (1): 39–48.
33. Dou, L., et al. 2024. Global natural hydrogen exploration and development situation and prospects in China (in Chinese). *Lithologic Reservoirs* 36 (02): 1–14.
34. HyTerra. 2023. Natural hydrogen exploration & production.
35. Hui, R., and A. Ding. 2018. Role of microorganisms in oil generation (II): Hydrogen metabolism and organic matter input from many origins (in Chinese). *Acta Sedimentologica Sinica* 36 (05): 1023–1031.
36. Leng, H., et al. 2020. Recent advances in hydrogenotrophic methanogenesis (in Chinese). *Acta Microbiologica Sinica* 60 (10): 2136–2160.
37. Zhang, X., H. Zhang, and L. Cheng. 2019. Key players involved in methanogenic degradation of organic compounds: Progress on the cultivation of syntrophic bacteria (in Chinese). *Acta Microbiologica Sinica* 59 (02): 211–223.
38. Zhang, J., and Y. Lu. 2015. A review of interspecies electron transfer insyntrophic-methanogenic associations (in Chinese). *Microbiology China* 42 (05): 920–927.
39. Zhou, F., et al. 2013. Formation mechanism of biogas in eastern Qaidam Basin (in Chinese). *Fault-Block Oil & Gas Field* 20 (04): 422–425.
40. Xia, Z., and Z. Bai. 2004. Discussion on CO_2 geological sequestration by methanogens in the biogenic gas field in China (in Chinese). *Petroleum Exploration and Development* 06: 72–74.
41. Shuai, Y., et al. 2010. The source of nutrient substrates for microbes in the deep biosphere and the characteristics of biogenic gas source rock. *Scientia Sinica (Terrae)* 40 (07): 866–872.
42. Shuai, Y., et al. 2009. Geochemical evidence for strong ongoing methanogenesis in Sanhu region of Qaidam Basin (in Chinese). *Scientia Sinica (Terrae)* 39 (06): 734–740.
43. Wu, J., et al. 2022. Influence of hydrogen fugacity on thermal transformation of sedimentary organic matter: Implications for hydrocarbon generation in the ultra-depth. *Scientia Sinica(Terrae)* 52 (11): 2275–2288.

9 Conclusion and Future Outlook

Natural hydrogen resource represents a transformative opportunity for clean energy production, originating through geologic mechanisms including serpentinization, radiolysis, microbial metabolism, and mantle-derived gas release. These processes operate within distinct geological environments characterized by ultramafic rocks, sedimentary basins, and Precambrian shield rocks, each contributing uniquely to H_2 accumulation. The development and preservation of these reservoirs depend on interconnected variables such as lithology, structural features, burial depth, pore networks, and hydraulic conductivity, combined with geochemical conditions like thermal gradients, fluid pressures, aqueous interactions, and reactive mineral assemblages. Biogenic contributions, specifically anaerobic microbial consortia, further modulate H_2 dynamics through redox-driven metabolisms. Exogenous influences such as crustal radiogenic decay, hydrothermal convection systems, and seismic faulting accelerate H_2 generation and subsurface transport. Prospects for natural hydrogen utilization are strengthening, propelled by growing recognition, financial commitments, and innovation in geoscientific methodologies. Cutting-edge advancements in subsurface imaging, spectral analysis, and targeted drilling techniques are improving exploration precision and operational viability. Governmental incentives and cross-sector collaborations are proving pivotal in establishing regulatory frameworks and scalable infrastructure. Meanwhile, declining production costs and scalable extraction methods are positioning H_2 as a competitive replacement for conventional hydrocarbons. The ecological and socioeconomic value of natural hydrogen is substantial, with the capacity to decarbonize industrial sectors and enable energy security through renewable-compatible storage. Synergistic integration with renewable infrastructure could optimize grid stability and mitigate variability challenges in green energy systems. As

Y. Yuan et al., *Natural Hydrogen*, Synthesis Lectures on Renewable Energy Technologies, https://doi.org/10.1007/978-3-032-14469-0_9

multidisciplinary research accelerates, natural hydrogen is anticipated to become a cornerstone of decarbonization strategies, driving global progress toward climate-resilient energy economies.

Natural hydrogen resource is widely distributed in diverse geological settings, including mid-ocean ridges, ophiolite complexes, Precambrian shields, volcanic zones, rift systems, and sedimentary basins. The primary mechanisms for hydrogen generation encompass serpentinization, radiolysis, mantle degassing, and microbial activity, highlighting its abiotic and renewable characteristics. This potential resource has attracted growing interest from numerous countries such as Australia, China, Mali, North America,, several EU nations, where exploratory research and extraction efforts have been initiated. Notably, the high-purity hydrogen field in Bourakebougou, Mali, has demonstrated the commercial viability of natural hydrogen extraction, providing a critical proof of concept for the global community. Although research on natural hydrogen in China started relatively late, several high-concentration hydrogen occurrences have been identified in geological settings highly conducive to hydrogen enrichment, indicating promising exploration potential. Favorable zones for natural hydrogen accumulation include the Tan-Lu Fault Zone and adjacent rift basins, the Altyn Tagh Fault Zone with its surrounding basins, as well as the Sanjiang Orogenic Belt–Longmenshan Fault Zone and associated basin systems.

However, the exploration and development of natural hydrogen resource face substantial challenges, primarily due to hydrogen's high mobility, low density, and strong reactivity, which complicate its accumulation and preservation. Effective trapping requires specific geological conditions, such as impermeable caprocks (e.g., evaporites or basaltic sills), structurally confined reservoirs, and reducing geochemical environments. Advances in detection technologies, including seismic and gas geochemical surveys, are essential to identifying and characterizing these elusive reservoirs. Moreover, the integration of multidisciplinary approaches, combining geology, geophysics, geochemistry, and engineering, will be vital to de-risking exploration and optimizing recovery strategies.

Looking ahead, the utilization of natural hydrogen is contingent upon overcoming technical and economic challenges associated with its extraction, storage, transportation, and safety. Future research efforts should focus on refining predictive models for hydrogen generation and migration processes, advancing cost-effective monitoring and extraction technologies, and comprehensively evaluating environmental impacts. Enhanced international collaboration and data sharing will be crucial to accelerate the understanding of natural hydrogen systems and facilitate the transition to a hydrogen-based economy. Through sustained innovation and investment, natural hydrogen holds transformative potential in supporting global decarbonization objectives and strengthening energy security.

9.1 Technical Challenges in Natural Hydrogen Development

Despite the promising potential of natural hydrogen, several technical and scientific challenges must be addressed to ensure its large-scale development [1–3]: (i) Exploration and Detection techniques: H_2 is a small and highly diffusive molecule, which makes it difficult to detect and trap compared to hydrocarbons. Current exploration techniques, such as seismic surveys and soil gas sampling, are being adapted, but there is a need for more refined tools and methods specifically for H_2. The lack of well-developed geological models for H_2 accumulation adds another layer of complexity to exploration efforts. (ii) Exploration Cost and benefits: H_2 reservoirs are still relatively new in the context of natural gas or oil exploration, so more advanced and costly technologies are required to accurately identify and assess them. (iii) Hydrogen Leakage and Containment: One of the most significant challenges in natural hydrogen development is ensuring that H_2 remains trapped in geological formations. As H_2 is highly mobile, and without secure geological traps, it can easily escape, reducing the amount available for extraction. (iv) Hydrogen can leak through even small fissures or faults in the subsurface rock, which requires robust trapping mechanisms, such as impermeable caprocks or geologically sealed reservoirs. Failure to contain H_2 can lead to significant losses. (v) Reservoir Characterization: Natural hydrogen are not yet as well understood as oil or gas fields. Key geological factors such as the size, quality, and rate of H_2 production within these reservoirs need further study. Without accurate reservoir characterization, predicting the economic viability of a given reservoir is challenging. (vi) Extraction Technology: While conventional drilling techniques can be used for H_2 extraction, the low density and high diffusivity of H_2 pose challenges. Existing infrastructure for natural gas might need modification to prevent leaks and ensure safe storage and transport of H_2 from wellheads to the market. (vii) Energy Yield and Economics: The energy yield from natural hydrogen, compared to synthetic H_2 production, is not yet fully understood. H_2 production from geological sources could be less expensive than electrolysis, but comprehensive economic analyses are still needed to determine the viability of this energy source on a commercial scale. (viii) Commercial Viability: While the extraction of H_2 from natural reservoirs could be cheaper than electrolyzing water, the commercialization of these sources requires significant infrastructure development, especially for transportation and storage.

9.2 Future Outlook

The future of natural hydrogen development looks promising, though its trajectory depends heavily on overcoming current technical challenges and increasing investment in research and infrastructure. Several factors indicate that natural hydrogen could play a pivotal role in the future energy landscape [5–10]: (i) Sustainability and environmental benefits: Unlike H_2 produced from natural gas (gray hydrogen) or through carbon-intensive processes, natural hydrogen can be produced with minimal CO_2 emissions. Its extraction could lead to a near-zero carbon energy supply, contributing significantly to decarbonization goals. (ii) Global energy demand: As countries aim to reduce their reliance on fossil fuels and meet climate targets, H_2 is seen as a critical energy carrier for industries such as heavy transportation, steel production, and power generation. If large natural hydrogen resource is economically developed, they could supplement or even replace the need for H_2 production through energy-intensive methods. (iii) Ongoing technological innovations: Breakthroughs in critical hydrogen technologies, including fuel cells, hydrogen storage, and transportation, will significantly influence the viability of natural hydrogen as a potential energy resource. Advances in drilling, reservoir management, and gas separation could help overcome many of the current technical limitations. (iv) Policy and market drivers: Government policies promoting renewable energy, carbon pricing, and international commitments to carbon neutrality could accelerate the development of natural hydrogen. As H_2 strategies become integral to national energy plans, the focus on natural hydrogen is expected to grow. (v) Potential for widespread deposits: Geological H_2 production is not limited to specific regions, and its generation could be continuous in areas with active tectonics. This opens the possibility of natural hydrogen being a globally distributed energy resource, providing countries without traditional hydrocarbon reserves with an alternative energy source.

Natural hydrogen development is still in its infancy, but it has the potential to transform the global energy system by providing a clean, low-cost, and sustainable source of hydrogen. Despite the technical challenges, such as difficulties in exploration, containment, and reservoir management, ongoing research and pilot projects are providing promising results.

In the long term, natural hydrogen is poised to emerge as a critical component of the hydrogen economy, contributing to global decarbonization initiatives by curbing greenhouse gas emissions and facilitating the shift from fossil fuel dependency. However, to fully realize this potential, significant investment in research, exploration technologies, and infrastructure is needed, along with collaborative efforts between governments, industries, and scientific communities.

References

1. Milkov, A. V. 2022. Molecular hydrogen in surface and subsurface natural gases: Abundance, origins and ideas for deliberate exploration. *Earth-Science Reviews* 230: 104063.
2. Smith, N. 2002. It's time for explorationists to take hydrogen more seriously. *First Break* 20 (4).
3. Zgonnik, V. 2020. The occurrence and geoscience of natural hydrogen: A comprehensive review. *Earth-Science Reviews* 203: 103140.
4. Epelle, E. I., et al. 2022. Perspectives and prospects of underground hydrogen storage and natural hydrogen. *Sustainable Energy & Fuels* 6 (14): 3324–3343.

5. Dou, L., et al. 2024. Global natural hydrogen exploration and development situation and prospects in China. *Lithologic Reservoirs* 36 (02): 1–14 (in Chinese).
6. Agency, I.E. 2023. *Net Zero Roadmap: A Global Pathway to Keep the 1.5 °C Goal in Reach.* IEA Paris, France.
7. Tian, Q., et al. 2022. Origin, discovery, exploration and development status and prospect of global natural hydrogen under the background of "carbon neutrality." *China Geology* 5 (4): 722–733.
8. Bendall, B. 2022. Current perspectives on natural hydrogen: A synopsis. *MESA Journal* 96: 37–46.
9. Osselin, F., et al. 2022. Orange hydrogen is the new green. *Nature Geoscience* 15 (10): 765–769.
10. Ellis, G.S., and A. Demas. 2023. *The Potential for Geologic Hydrogen for Next-Generation Energy.* US Geological Survey.

The manufacturer's authorised representative in the EU is Springer Nature Customer Service Centre GmbH, Europaplatz 3, 69115 Heidelberg, Germany. If you have any concerns regarding our products, please contact ProductSafety@springernature.com

Printed and bound by CPI Group (UK) Ltd, Croydon, CR0 4YY
07/07/2026
02160919-0003